ACKNOWLEDGEMENT

- Feedback™ is a trademark of Feedback Corporation.

D1303979

PREFACE

This is a textbook used for the junior/senior undergraduate students who majored in Computer Engineering, Electrical Engineering or Electronics Technology.

Prior to take this class, students are supposed to have finished the fundamental classes in electronics and electrical engineering, such as the electronics circuit design and analysis, network analysis and linear circuit analysis.

This textbook is a lab oriented and all contents in this book is composed by a sequence of lab experiments, from the basic DC/AC, resistor/capacitor/inductor networks, diodes and transistors, amplifiers, OP integrated amplifier to all filters. The labs in this book is organized based on the components taught in the general electronics textbook, from top to bottom, from surface to deeper point, to show students how to design and implement different components to build circuit units to meet the different requirement in the real applications in this world.

A Feedback lab toolkit Basic Electricity Kit EEC471, which is provided by Feedback™ corporation, is used through this textbook. If this toolkit is not available, one can use the general-purpose breadboard to replace this kit without any problem.

The first section, Warmup, which is a review process, can be skipped if students had a solid understanding of the fundamental electronics components and principles. The lab 9 is an A/D converter that belongs to a higher level application of the interface between the digital circuits and analog circuits. Although some additional digital circuits knowledge is required for this lab, the author still keep this lab in this book since the A/D converter is a very popular device used in the analog and digital circuits design and applications.

Each lab contains a sequence of questions that ask students to consider and answer in the lab report. Those can be considered as a criteria and used to evaluate the students' degree to which they have understood and mastered for this class.

Upon finish this lab sequence, students can have both solid theoretical knowledge and hand-on experimental experience in electronics circuit design and implementation.

CONTENT

LAB 0
BASIC ELECTRONICS COMPONENTS

❖ PREFACE

- ➤ Atoms, Nucleus, Protons, Neutrons and Electrons.
- ➤ Conductors, Insulators and Semi-conductors.
- ➤ Potential, Potential Difference and Voltage.
- ➤ Definite Movement of Electrons – Current.
- ➤ Work and Power.
- ➤ DC and AC.

❖ BASIC COMPONENTS AND CIRCUITS

- ➤ Resistor and Color Code
- ➤ Series in Resistors
- ➤ Parallel in Resistors
- ➤ Series-Parallel in Resistors (Simple Network)
- ➤ Superposition Theorem
- ➤ Thevenin's Theorem
- ➤ Simple Network Calculation

  - ❑ Loop Current Method
  - ❑ Node Voltage Method

○ Neutrons
⊕ Protons
⊖ Electrons

❖ WARM UP

➤ Atoms, Nucleus, Protons, Neutrons and Electrons

- • Atoms – Composed of Nucleus and Electrons.
- • Nucleus – Composed of Protons and Neutrons.
- • Protons are Positive charged, and Electrons are Negative charged. The number of the Protons is always equal to the number of the Electrons.
- • The Electrons always circle around the Nucleus in orbits.
- • There is a force of attraction between the Protons and Electrons because the Protons are positive charged and Electrons are Negative charged. However, the Electrons can not be attached to the Nucleus by this Positive force because of the Centrifugal forces created by the rotating movement of the Electrons around the Nucleus. That is the balance of the movement.

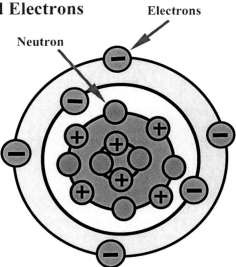

Orbital Electrons

Neutron

➢ An Atom is electrically neutral because the number of the Protons and the number of the Electrons is EQUAL

➢ Conductors, Insulators and Semiconductors

Electrons have a random movement in the Atoms.

- **Conductors** – Most Electrons are easily moved if a potential difference is applied to these Materials. The moving direction of the Electrons is from the high Potential position to the low Potential position.

- **Insulators** - Less or almost no Electrons are easily moved.

  Conductors are called good conductor, and Insulators are called bad conductors.
  Most metal materials are good conductors. Most insulating medium are good insulator (Wood, Paper, Rubber and Plastics)

- **Semiconductors** – The conductivity between Conductors and Insulators are called Semiconductors (Silicon, Germanium).

➢ Potential, Potential Difference and Voltage

<u>Electrical Potential</u> is a something like an electrical level or electrical position. Such as a 6V battery, the potential at its positive terminal is 6V (Volts), and the potential at its negative terminal is 0V if that terminal was selected as a reference position.

Potential = 0 V
Potential = 6 V

This is analogous to the definition of the gravitational potential energy through the work done by the force of gravity in moving a mass through a certain distance. The units of potential are Joules / Coulomb, which are called **Volts** (V).
Potential is a relative definition, no absolute potential position.

<u>Electrical Potential Difference</u> is the difference between the two Potential levels.

Physically, potential difference has to do with how much work the electric field does in moving a charge from one place to another. Batteries, for example, are rated by the potential difference across their terminals. In a nine volt battery the potential difference between the positive and negative terminals is precisely nine volts. On the other hand the potential difference across the power outlet in the wall of your home is 110 volts.

Potential difference is also called Voltage or Voltage drop across a conductor. The unit of the potential difference is also Volts, same as the potential unit.

➢ Current – Electrons Definite Movement

If a Potential difference (Voltage) is applied on a conductor, the Electrons (Charge) will be forced to move from the high potential position (Positive terminal) to the low potential position (Negative terminal).

The motion of the Positive Charge in a certain direction (from positive terminal to Negative terminal) creates: **Current**.

• The Direction of the **Current** is defined by the moving direction of the Positive Charge. That is, the positive charge moves from the Positive terminal to the Negative terminal.

In the real world, **<u>No any Positive charge existed at all!</u>**
The only Charge is **<u>Negative Charge</u>**. The real current Direction is the moving direction of Electrons, that is, the Electronics move from Negative terminal to the Positive terminal.

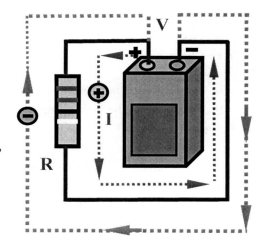

But because the human being used to use the positive charge, So this definition has been used until today.

All following teaching materials for this class still use the conventional definition, that is:
Current direction is the moving direction of the positive charge in the circuit.

Current: $I = \dfrac{q}{t}$; The mount of charges move through an intersection within a unit time period.

➢ Work and Power

Inside Circuit and External Circuit:
The Circuit inside the battery (power supply) is called **internal circuit**.
The circuit outside of the battery (power supply) is called **external circuit**.

In external circuit, the Positive Charge moves from
the high potential to lower potential to form the
Current flowing. There must be another 'Force'
to move the charge from the lower potential to
higher potential inside the battery (inside circuit)
in order to keep the current flow
continuously.

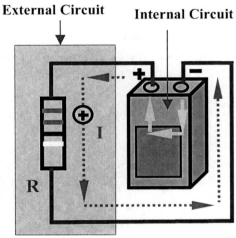

This so-called force is called the **electromotive force**,
or **emf**. The SI unit for the **emf** is a volt (and thus this
is not really a force, despite its name). We will use a
script **E** to represent the **emf**.

When emf moves a unit charge from lower potential to the
higher potential in the inside circuit, a Work is done. The Work done at a unit time is called ower,
that is

Work: The amount of the energy when moves a unit charge from a potential difference

$$U = V_a - V_b$$

$$W = q \times U$$

$$P = \frac{W}{t} = \frac{q \times U}{t} = I \times U; \qquad \text{so:} \qquad P = I \times U;$$

➢ DC and AC

Direct Current Circuit: The Current is a <u>Stable</u> value, and doesn't change with the time.

Alternative Current Circuit: The Current is a <u>Periodical Function</u> of the time.

Components used in DC circuits:

Resistor, DC Lamps, LED (Light Emitted Diode)

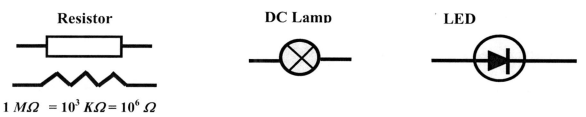

Resistor   **DC Lamp**   **LED**

$1 \, M\Omega = 10^3 \, K\Omega = 10^6 \, \Omega$

Components used in AC circuits:

Capacitor, Inductor, Transformer, AC Lamps

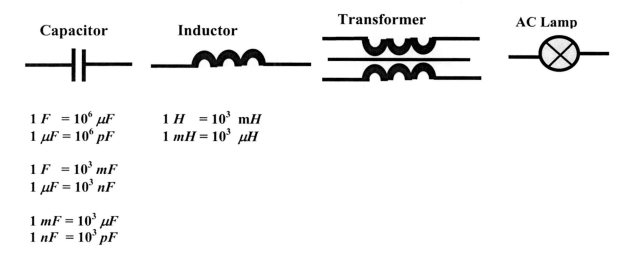

Capacitor   **Inductor**   **Transformer**   **AC Lamp**

$1 \, F = 10^6 \, \mu F$
$1 \, \mu F = 10^6 \, pF$

$1 \, H = 10^3 \, mH$
$1 \, mH = 10^3 \, \mu H$

$1 \, F = 10^3 \, mF$
$1 \, \mu F = 10^3 \, nF$

$1 \, mF = 10^3 \, \mu F$
$1 \, nF = 10^3 \, pF$

❖ BASIC COMPONENTS AND CIRCUITS

➤ Resistor and Color Code

Resistors Can provide resistance to the electronic circuits.

Resistors are electronic components used extensively on the circuit boards of electronic equipment. They are color coded with stripes (Bands) to reveal their resistance value (in ohms) and their manufacturing tolerance.

Resistors are Linear Components in the circuits. That means the Resistor's value, Resistance, doesn't change with the time, frequency of the circuits.

Resistor' Unit Ohms (Ω), Kilo Ohms ($K\Omega$) and Mega Ohms ($M\Omega$).

Resistors can be used to work as a Voltage divider (Series connection), Current divider (Parallel connection) and network in the circuits design.

Calculation of a Conductor's Resistor:

$$R = \rho \, \frac{L}{S};$$

ρ = Resistor coefficient,

L = Conductor's Length,

S = Conductor's Intersectional Area.

Color Code

Resistor Color Codes

Color codes are used to mark the value of most small resistors. The most common color code is the three or four band method described in this exercise. The first band is a code for the first digit. The second band is a code for the second digit. The third band is a multiplier in powers of ten. The fourth band is a code for the tolerance or accuracy of the resistor. Color codes are not used on all resistors. Large resistors often have values printed on them. The chart below explains the meaning of each resistor color band.

First Digit
Second Digit
Tolerance
Multiplier

Color	First Digit	Second Digit	Multiplier
Black	0	0	1
Brown	1	1	10
Red	2	2	100
Orange	3	3	1000
Yellow	4	4	10000
Green	5	5	100000
Blue	6	6	1000000
Violet	7	7	10000000
Grey	8	8	100000000
White	9	9	1000000000

Tolerance Color	Error Range
None	20%
Silver	10%
Gold	5%

Note:

The multiplier band can also be silver or gold. This is used to place values of less than one ohm on resistors.

Silver Multiplier Band = 0.1
Gold Multiplier Band = 0.01

Reading resistor color codes is a useful if not essential skill for those who work with electronic circuits. This exercise will present you with a series of resistor color code problems to solve. The problems will be presented in random order.

Begin

Quit

Tue. 28-Aug-2001 09:29:10

Ohms Laws

Ohm's Law One

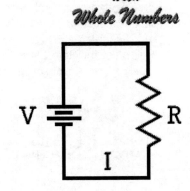

Ohm's Law with Whole Numbers

Three important quantities in electrical circuits are voltage, current, and resistance. The relationship between these three quantities is known as Ohm's law. Ohm's law can most easily be expressed in mathematical form as shown below. The letter, V, is used to represent voltage measured in volts. The letter, R, represents resistance measured in ohms. The letter, I, represents current flow measured in amperes. If any two of the three quantities are known, the third can be calculated by use of the proper relationship. Simply pick the relationship that has the known quantities on the right and the unknown on the left side of the equal sign.

$$V = IR \qquad I = \frac{V}{R} \qquad R = \frac{V}{I}$$

Anyone working with electricity is expected to be able to apply Ohm's law. This exercise will present you with a series of problems to solve using Ohm's law. Be aware that the problem order and quantities will be different each time you go through this exercise. Move your mouse pointer to a formula to obtain additional hints. Your answers are graded for 5% or better accuracy.

Begin

Quit

Tue. 28-Aug-2001 14:45:10

- This page is copyrighted by **ETCAI Products Inc.**

Power Equations

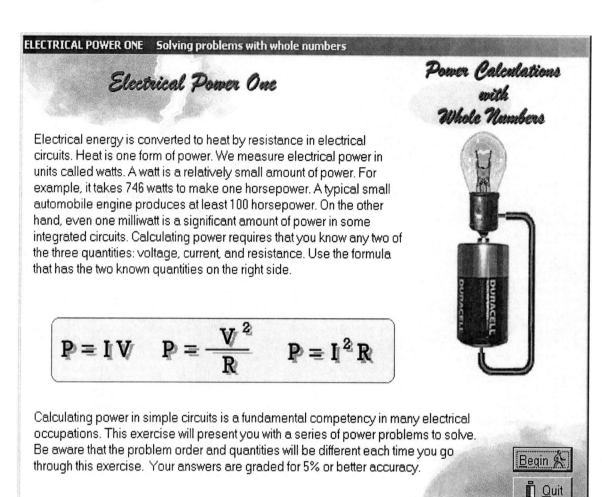

Electrical Power One

Power Calculations with Whole Numbers

Electrical energy is converted to heat by resistance in electrical circuits. Heat is one form of power. We measure electrical power in units called watts. A watt is a relatively small amount of power. For example, it takes 746 watts to make one horsepower. A typical small automobile engine produces at least 100 horsepower. On the other hand, even one milliwatt is a significant amount of power in some integrated circuits. Calculating power requires that you know any two of the three quantities: voltage, current, and resistance. Use the formula that has the two known quantities on the right side.

$$P = I V \qquad P = \frac{V^2}{R} \qquad P = I^2 R$$

Calculating power in simple circuits is a fundamental competency in many electrical occupations. This exercise will present you with a series of power problems to solve. Be aware that the problem order and quantities will be different each time you go through this exercise. Your answers are graded for 5% or better accuracy.

Begin

Quit

Tue 28-Aug-2001 14:47:31

- This page is copyrighted by **ETCAI Products Inc.**

Resistor in Series and Parallel Connection

- SERIES CIRCUIT

$$U = U_1 + U_2 + U_3 = IR_1 + IR_2 + IR_3 = I(R_1 + R_2 + R_3)$$

$$= IR$$

$$R = R_1 + R_2 + R_3$$

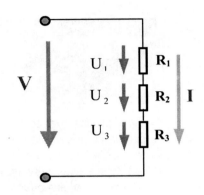

-PARALLEL CIRCUIT

$$I = I_1 + I_2 + I_3 = \frac{U}{R_1} + \frac{U}{R_2} + \frac{U}{R_3} = U\left(\frac{1}{R_1} + \frac{1}{R_2} + \frac{1}{R_3}\right) = \frac{U}{R}$$

$$\frac{1}{R} = \frac{1}{R_1} + \frac{1}{R_2} + \frac{1}{R_3}$$

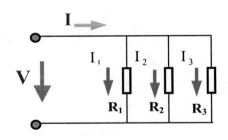

Reference Guide to Laboratory Instruments

This section is provided to help familiarize you with basic laboratory instruments and may be used as a reference as the instruments are introduced in the experiments. It is impossible to cover all possible variations between instruments so only general features, common to a class of instruments, are described. Consult the operator's manual for detailed descriptions and safe operating practice of the particular instruments in your laboratory.

The Power Supply

Most electronic circuits require a source of regulated direct current (dc) to operate properly. A direct current regulated power supply is a circuit that provides the energy to allow electronic circuits to function. They do this by transforming a source of input electrical power (generally ac) into dc. Most regulated supplies are designed to maintain a fixed voltage that will stay within certain limits of voltage for normal operation. Voltage adjustment and current limits depend on the particular supply.

The power supply must provide the proper level of dc voltage for a given circuit. Some integrated circuits, for example, can function properly only if the voltage is within a very narrow range. You will normally have to set the voltage to the proper level before you connect a power supply to the test circuit. The power supply at your bench may have more than one output and normally will have a built- in meter to help you set the voltage. Some power supplies have meters that monitor both voltage and current. There may be more than one range or several supplies built into the same chassis, so the meter may have multiple or complex scales.

It is important that the user make good connections to the power supply output terminals with wire that is sufficient to carry the load current if the output were accidentally shorted together. Clip-leads are not recommended as they can produce measurement error due to high contact resistance. In situations where several circuits are operated from the same supply, the best policy is to operate each circuit with an independent set of leads.

The Multimeter

The digital multimeter (DMM) and analog volt-ohm-milliammeter (VOM) are multipurpose measuring instruments that combine the characteristics of a dc and ac voltmeter, dc and ac ammeter, and an ohmmeter in one instrument. The DMM indicates the measured quantity as a digital number, avoiding the necessity to interpret the scales as is required on analog instruments. Although in most labs the DMM has replaced the VOM as the instrument of choice, there are several advantages to the VOM. It is less susceptible to interference and has a much higher frequency response.

Because the multimeter is a multipurpose instrument, it is necessary to determine which controls select the desired function. In addition, current measurements (and often high-range voltage measurements) usually require a separate set of lead connections to the meter. After you have selected the function, you need to select the appropriate range to make the measurement. It is important to select the function and range *before* connecting the meter to the circuit you are testing. DMMs can be autoranging, meaning that the instrument automatically selects the correct scale and sets the decimal place; or they can be manual ranging, meaning that the user must select the correct scale. For manual ranging instruments, when the approximate voltage or current is not known, always begin a measurement on the highest possible range to avoid instrument overload and possible damage. Change to a lower range as necessary to increase the precision. The life of range switches will be lengthened if you only change ranges with the probes disconnected from the circuit. On analog instruments the range selected should give a reading in the upper portion of the scale.

The voltmeter function of a DMM can measure either ac or dc volts. The dc voltage function is useful to measure the dc voltage *difference* between two points. If the meter's red lead is touching a more positive point than the meter's black lead, the reading on the meter will be positive; if the black lead is on the more positive point, the reading will be negative. Analog meters *must* be connected with the correct polarity, or the pointer will attempt to move backward, possibly damaging the movement.

The ac voltage function is designed to measure low-frequency sinusoidal waveforms. The meter is designed to indicate the rms (root-mean-square) value of a sinusoidal waveform. Frequency is the number of cycles per second, measured in Hz, for a waveform. All DMMs and VOMs are limited to some specified frequency range. The meter reading will be inaccurate if you attempt to measure waveforms outside the meter's specified frequency range. A typical DMM is not accurate on the ac scale below about 45 Hz or above about 1 kHz, although this range can be considerably better in some cases. A VOM can measure ac waveforms over a much larger range up to 100 kHz.

The ohms function (used for resistance measurements) is used only in circuits that are *not* powered. An ohmmeter works by inserting a small test voltage into a circuit and measuring the resulting current. Consequently, if any voltage is present, the reading will be in error. The meter will show the resistance of all possible paths between the probes. If you want to know the resistance of a single component, it is necessary to isolate that component from the remainder of the circuit by disconnecting one end. In addition, body resistance can affect the reading if you are holding the conducting portion of both probes in your fingers. This procedure should be avoided, particularly with high resistances.

The Function Generator
The basic function generator is used to produce sine, square, and triangle waveforms and may also have a pulse output for testing digital logic circuits. Function generators normally have certain controls that allow you to select the type of waveform and other controls to adjust the amplitude and dc level. The peak-to-peak voltage is adjusted by the amplitude control. The dc level is adjusted by a control labeled dc offset; this enables you to add or subtract a dc component to the waveform. These controls are generally not calibrated, so amplitude and dc level settings need to be verified with the oscilloscope or multimeter.

The frequency is selected with a combination of a range switch and vernier control. The range is selected by a decade frequency switch or pushbuttons that enable you to select the frequency in decade increments (factors of ten) up to about 1 MHz. The vernier control is usually a multiplier dial for adjusting the precise frequency needed.

The output level of a function generator will drop from its open circuit voltage when it is connected to a circuit. Depending on the conditions, you generally will need to readjust the amplitude level of the generator after it is connected to the circuit. This is because of the generator's Thevenin resistance that will affect the circuit under test. Common values of Thevenin resistance are 50 and 600.

Higher-priced instruments will add features such as trigger or sync outputs to use in synchronizing an oscilloscope, modulation, increased frequency ranges, fixed attenuators on the output, and so forth. Fixed attenuators are handy if you want to reduce the output by an exact amount or you want to choose a very small, but known signal. Some function generators have a symmetry or duty cycle control that allows you to control the pulse width of the rectangular pulse. Details of the particular features of your function generator and the controls can be found in the operator's manual.

The Oscilloscope

The oscilloscope is the most versatile general-purpose measuring instrument, letting you see a graph of the voltage as a function of time in a circuit. Many circuits have specific timing requirements or phase relationships that can be readily measured with a two-channel oscilloscope. The voltage to be measured is converted into a visible display by a cathode ray tube (CRT), a vacuum device similar to a television picture tube.

The oscilloscope contains four functional blocks. The input signal is connected to the **vertical** section which can be set to attenuate or amplify the input signal to provide the proper voltage level to the vertical deflection plates of the CRT. The **trigger** section samples the input waveform and sends a synchronizing trigger signal at the proper time to the horizontal section. The trigger occurs at the same relative time to superimpose each succeeding trace on the previous trace. This action causes the signal to appear to stop, allowing you to examine the signal. The **horizontal** section contains the time-base (or *sweep*) generator which produces a linear ramp or ``sawtooth" waveform that controls the rate the beam moves across the screen. The horizontal position of the beam is proportional to the elapsed time from the start of the sweep, allowing the horizontal axis to be calibrated in units of time. For this reason, the horizontal section is often called the time base. The output of the horizontal section is applied to the horizontal deflection plates of the CRT.

Finally, the **display** section contains the CRT and beam controls. It enables the user to obtain a sharp presentation with the proper intensity. The display section frequently contains other features. Sometimes, the user can lose the displayed waveform by accidentally positioning it offscreen. One common feature in the display section is a beam finder button that enables the user to quickly locate the position of the trace. Controls for each of the functional blocks are usually grouped together. Frequently, there are color clues to help you identify groups of controls. Details of these controls are explained in the operator's manual for the oscilloscope; however, a brief description of frequently used controls is given in the following paragraphs.

Display Controls

The display system contains controls for adjusting the electron beam. The focus and intensity controls should be adjusted for a comfortable viewing level with a sharp focus. The display section may also contain the beam finder, a control which is used in combination with the horizontal and vertical position controls to bring the trace on the screen. Another control over the beam intensity is the *z*-axis input. A control voltage on the *z*-axis input can be used to turn on or off the beam or adjust its brightness. Some oscilloscopes also include the trace rotation control in the display section. Trace rotation is used to align the sweep with a horizontal gratitude line. This control is usually adjusted with a screwdriver to avoid accidental adjustment.

Vertical Controls

The vertical controls include the volts/div (vertical sensitivity) control and its vernier, the input coupling switch, and the vertical position control. Multiple-channel oscilloscopes will have a duplicate set of these controls for each channel and various switches for selecting channels or other vertical operating modes. The vertical input is connected through a selectable attenuator to a high input impedance dc amplifier. The volts/div control selects a combination of attenuation/gain to determine the vertical sensitivity of the scope. For example, a low-level signal will need more gain/less attenuation than a higher level signal. The vertical sensitivity is adjusted in fixed volts/div increments to allow the user to make calibrated voltage measurements. In addition, a concentric vernier control is usually provided to allow a continuous range of sensitivity. This knob must be in the detent (calibrated) position to make voltage measurements. The detent position can be felt by the user as the knob is turned because the knob tends to ``lock" in the detent position. Some oscilloscopes have a warning light or message when the vernier is not in its detent position.

The input coupling switch is a multiple-position switch that can be set for ac-gnd-dc and sometimes includes a 50 position. The gnd position of the switch internally *disconnects* the signal from the scope and grounds the input amplifier. This position is useful if you want to set a ground reference level on the screen for measuring the dc component of a waveform. The ac and dc positions are high impedance inputs typically 1 M shunted by 15 pF of capacitance. High impedance inputs are useful for general probing at frequencies below about 1 MHz. At higher frequencies, the shunt capacitance can load the signal source excessively, causing measurement error. Attenuating divider probes are good for high-frequency probing because they have very high impedance (typically 10 M) with very low shunt capacitance (as low as 2.5 pF).

The ac position of the coupling switch inserts a series capacitor before the input attenuator, causing dc components of the signal to be blocked. This position is useful if you want to measure a small ac signal riding on top of a large dc signal power supply ripple for example. The dc position is used when you want to view *both* the ac and dc components of a signal. This position is best when viewing digital signals as the input *RC* circuit forms a differentiating circuit. The ac position can distort the digital waveform because of this differentiating circuit. The 50 position places an accurate 50 load to ground. This position provides the proper termination for probing in 50 systems and reduces the effect of a variable load which can occur in high impedance termination. The effect of source loading *must* be taken into account when using a 50 input. It is important not to overload the 50 input as the resistor is normally rated for only 2 watts implying a maximum of 10 V rms of signal can be applied to the input.

LAB 1

DC NETWORK

Section A. Ohms Law

Objective

To explore the idea of the resistance of a component.

Equipment Required

EEC470 Constructor

Basic Electricity Kit EEC471

Power Supply Unit – 0~20V variable DC regulated (Feedback Power Supply PS445)

Multimeter and Ammeter 0-100 mA DC

Procedure

Make circuit connection as Fig. 2.3 below.

Connect the power supply unit to the main supply line.

Don't switch on yet.

Turn the variable DC control knob to minimum.

Lab Report Due

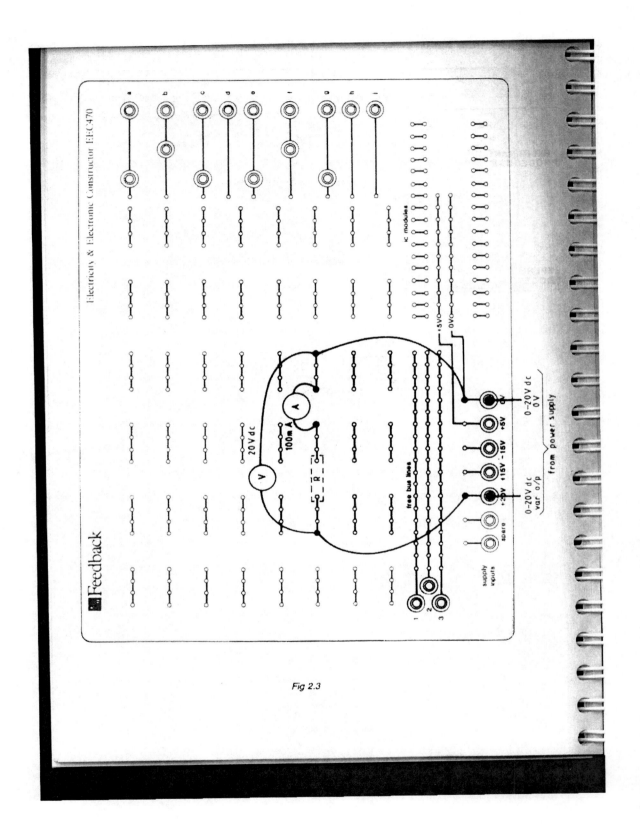

Fig 2.3

First construct the circuit as shown in the patching diagram of fig 2.3
The circuit for the above connection is given in fig 2.4

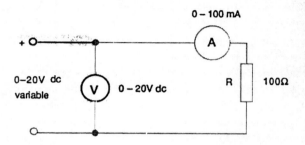

Fig 2.4

Set the meter to monitor the variable d.c voltage.

Make sure that the variable d.c control knob is fully counterclockwise, and then switch on the power supply..

We have noticed how the current changes from resistor to resistor when a fixed emf is applied, now let us see if there is any recognisable relationship between current and voltage for each of the resistors in turn.

Increase the applied voltage in 1V steps from 0V up to 10V, and at each step measure the current flowing in the resistor, as shown on the ammeter.

Copy the table of current against voltage as shown in fig 2.5, reproduced at the end of this assignment and note your results. Then plot a graph of V against I as shown in fig 2.6

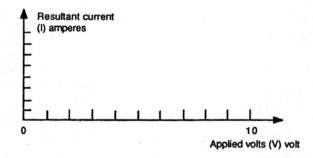

Fig. 2.6

17

1 What is the shape of the graph you obtain?
2 With this shape of graph, can you come to any conclusions as to the relationship between current and voltage for the resistor ?

Calculate the slope of the graph as outlined in fig 2.7

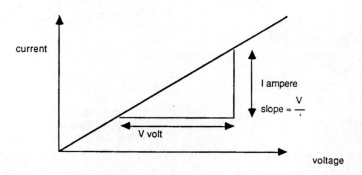

Fig 2.7

Now turn the voltage back to zero, and replace the 100Ω resistor with one of 1kΩ, as shown in fig 2.8.

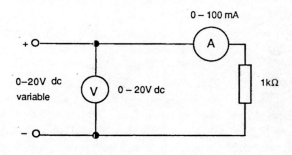

Fig 2.8

Now repeat the procedure making a table of voltage and current, and plotting a graph as before.

Question 3: Is the graph the same shape as before?

Question 4: What is the slope of this second graph? (Note: Always work out the slope with **V** in volts and **I** in amperes, not million-amperes).

Question 5: Do you then think that the ratio of voltage to current in a resistor, at D-C, could be used as a measure of its resistance?

We have seen that the unit of resistance called the OHM, named after George Simon Ohm who first discovered the relationship we have just investigated, is determined by the ratio of voltage to current. Often the word 'ohm' is given the symbol Ω (the Greek letter Omega).

i.e 20 ohms = 20 Ω

1000 ohms = 1 kΩ.

Question 6: How is the value of a 4700 Ω resistor written using this notation?

Question 7: Are the values measured for the two resistors within the tolerance given by their colored bands?

Question 8: What is the error between the real measured resistance and the nominal value? What is the percentage representation for this error?

Question 9: Where these errors come from?

Question 10: Calculate the mean value and standard deviation for both measured resistor values based on **Table 1** on next page.

For the mean value, using measured resistor values in **Table 1**:

$$M = \frac{R_1 + R_2 + \ldots\ldots + R_{10}}{10}$$

Standard Deviation: $\sigma = \sqrt{\sum_{i=1}^{10} \frac{(R_i - M)^2}{10}}$ $\qquad$ ($R_i = R_{imeas}$)

Question 11: What is the meaning of the Mean? What is the meaning of the Standard deviation?

Question 12: Is the error of measured resistor value acceptable (within the tolerance)?

Table 1.

MEASURED RESULT

Applied Volts (V)	Measured Current (mA)	Measured Resistor (Ω)	Nominal Resistor (Ω)	Error Value (Ω)
1				
2				
3				
4				
5				
6				
7				
8				
9				
10				

* When calculate the resistor values by using the measured currents, use Ampere, NOT mA value.

Collect answers to the questions above with plotting and result table together. That is your Lab report for Section 1.

Section B. Resistor Network

Objective

To investigate what happens when resistors are interconnected in a circuit.

Equipment Required

EEC470 Constructor

Basic Electricity Kit EEC471

Power Supply Unit – 0~20V variable DC regulated (Feedback Power Supply PS445)

Multimeter and Ammeter 0-10 mA DC

Procedure

Make circuit connection as Fig. 3.2 below.

Connect the power supply unit to the main supply line.

Don't switch on yet.

Turn the variable DC control knob to minimum.

Lab Report Due

Connect the power supply unit to the mains supply line. Ensure that the variable d.c control is at minimum. DO NOT switch on yet.

Let us now investigate the currents and voltages present when several resistors are connected together to provide a network of resistors.

Connect up the circuit as shown in the layout of fig 3.1

The circuit for the above connection is given in fig 3.2.

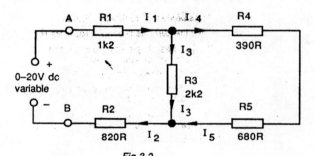

Fig 3.2

First ensure that the variable d.c control knob is fully counterclockwise, then switch on the psu. Adjust its output voltage to be 20V.

We wish to investigate the currents I, and the voltages across, each branch of the network. Let us first measure the voltages.

Using the 0-10V voltmeter, measure the voltage across R_1. Note the polarity of the voltage. Repeat the measurement for each of the other resistors.

Copy the table as shown in fig 3.3, reproduced at the end of this assignment, and tabulate your results . Also draw a circuit diagram of the network , as shown in fig 3.4, and mark the voltages on it, with their polarities.

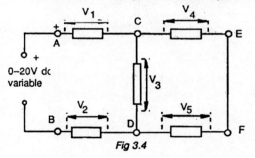

Fig 3.4

22

Summarising the two laws mathematically we may state:

Kirchhoff's current law: $\sum I = 0$

Kirchhoff's voltage law: $\sum V = 0$

Using these laws, the currents and voltages can be theoretically calculated without having to do measurements.

If we do not know what the voltage or current is going to be at the start of a theoretical calculation, we must assume that it is one particular way round, i.e assume its polarity, and take that assumed polarity into account. If our assumption was wrong, the answer will simply come out negative.

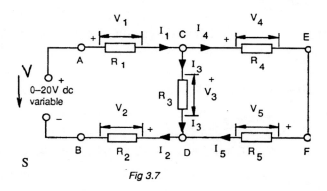

Fig 3.7

Referring to fig 3.7

For loop ACBD

$\sum V = 0$,   i.e $V_1 + V_2 + V_3 - V = 0$

Now $V_1 = I_1 R_1$; $V_2 = I_2 R_2$; $V_3 = I_3 R_3$

$\therefore I_1 R_1 + I_2 R_2 + I_3 R_3 - 20 = 0$ (1)

For loop CEFD

$\sum V = 0$;   i.e $V_4 + V_5 - V_3 = 0$

23

Now $V_3 = I_3 R_3$; $V_4 = I_4 R_4$; $V_5 = I_5 R_5$

So: $I_4 R_4 + I_5 R_5 - I_3 R_3 = 0$ (2)

Similar to loop ACEFDB,

$I_1 R_1 + I_4 R_4 + I_5 R_5 + I_2 R_2 - 20 = 0$,

but this is a redundant equation, telling us nothing further, since we could obtain it by adding together equations (1) and (2).

FOR NODE C

$\sum I = 0$, so $I_1 = I_3 + I_4$ (3)

For node D:

$\sum I = 0$ so $I_2 = I_3 + I_5$ (4)

A fifth equation is needed to determine the five unknown currents. In this case it is fairly obvious that :

$I_1 = I_2$ (they are the same current)   (5)

And $I_4 = I_5$ similarly

Equations 1 to 5 enable the currents to be determined. Using (5) we can write I_4 instead of I_5 in equation (4).

$I_2 = I_3 + I_4$ (6)

Combining equations (3) and (6) we deduce that:

$I_1 = I_2$ (7)

Equations (3), (5) and (6) enable all the currents to be expressed in terms of I_3 and I_4.

Applying this to equations (1) and (2), they become:

$(I_3 + I_4)(R_1 + R_2) + I_3 R_3 - 20 = 0$ (8)

$I_4 (R_4 + R_5) - I_3 R_3 = 0$ (9)

Equation (8) can be rewritten:

$I_4 (R_1 + R_2) + I_3 (R_1 + R_2 + R_3) - 20 = 0$ (10)

Substituting the known resistance values in equation (9) gives:

$I_4(390 + 680) - I_3(2200) = 0.$

So: $1070\,I_4 - 2200\,I_3 = 0$ (11)

and in equation (10) gives

$I_4(1200 + 820) + I_3(1200 + 820 + 2200) - 20 = 0.$

So: $2020\,I_4 + 4220\,I_3 = 20$ (12)

From equation (11)

$$I_3 = \frac{1070}{2200} \times I_4$$

Substituting in equation (12)

$$2020\,I_4 + \frac{4220 \times 1070}{2200} \times I_4 = 20$$

So: $I_4 = \dfrac{20}{4072} = 0.00491\ A = 4.91\ mA.$

From (11), $I_3 = 2.39\,mA.$

From (6), $I_2 = 2.39 + 4.91 = 7.30\ mA.$

(5) and (7) complete the determination of the currents. All the answers are positive, so we chose the directions of the arrows correctly.

Table 2.

Resistor Branch	Voltage (V)
R_1	
R_2	
R_3	
R_4	

Table 3.

Resistor Branch	Current I (mA)
R_1	
R_2	
R_3	
R_4	
R_5	

Table 4.

Resistor Branch	Marked Value Ω	Current I (mA)	Voltage V (V)	Actual Value Ω
R_1	1200			
R_2	820			
R_3	2200			
R_4	390			
R_5	680			

Question 1: With reference to Fig. 3.4, can you notice any relationship between the voltages round the loop ACDB? (Remember the polarities).

Question 2: With reference to Fig. 3.4, can you notice any relationship between the voltages round the loop CEFD? (Remember the polarities).

Question 3: With reference to Fig. 3.4, can you notice any relationship between the voltages round the loop AEFB? (Remember the polarities).

Question 4: With reference to Fig. 3.7, can you notice any relationship between the currents at the node C and D? (Remember the polarities).

Question 5: With reference to Fig. 3.7, what is the meaning if a current value is negative value (such as $I_s = -2$ A)?

Question 6: Are the directions of the currents shown in Fig. 3.2 correct?

Question 7: What can you say about the currents I_1, I_3 and I_4 at node C?

Question 8: Does the same apply for current I_2, I_3 and I_5 at node D?

Question 9: What is the algebraic sum of the voltages around a loop in a circuit?

Question 10: What is the algebraic sum of the currents at a node in a circuit?

Question 11: Compare the theoretical values of voltages and currents for each resistor with the measured results. Does the latter match to the former? What are the differences and why we have these differences?

Question 12: By what percentage do the actual values differ from the nominal (marked) value? ($\dfrac{Actual\ Value - No\min al\ Value}{No\min al\ Value} \times 100$)

LAB 2

— SECTION A

SUPERPOSITION THEOREM

SUPERPOSITION THEOREM

OBJECTIVE

To investigate the effects of more than one voltage source in a network.

EQUIPMENT REQUIRED

Qty	Apparatus
1	Electricity & Electronics Constructor EEC470
1	Basic Electricity Kit EEC471
1	Power supply unit
	0 to +20V variable dc regulated and
	+ 15V dc regulated
	(eg Feedback Power Supply PS445)
2	Multimeters or
1	Voltmeter 0-20V dc and
1	Milliammeter 0 – 10 mA dc

PREREQUISITE ASSIGNMENTS

KNOWLEDGE LEVEL

See prerequisite assignments

Lab Report Due:

Experiment Circuit

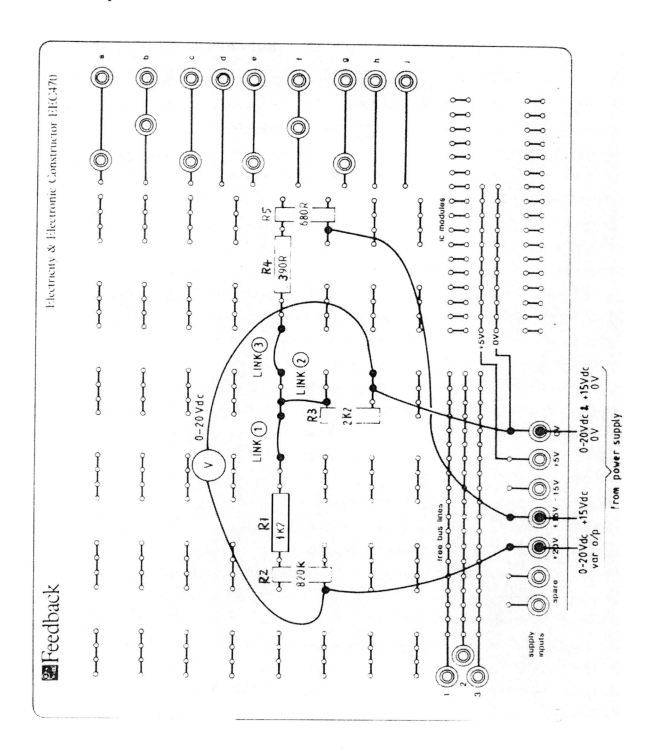

Figure 5.2

Experiment Procedure

Connect the layout as shown in fig 5.2 and check with the circuit diagram of fig 5.1.

Connect the power supply unit to the mains supply line.

Ensure that the variable d.c control is at minimum. DO NOT switch on yet. .

EXPERIMENTAL PROCEDURE

Previous experiments have shown what currents and voltages are present in a resistive network with one source of emf. We now wish to investigate networks which have more than one source, and to try to formulate some expressions for the resultant currents.

Let us investigate the network shown in fig 5.1.

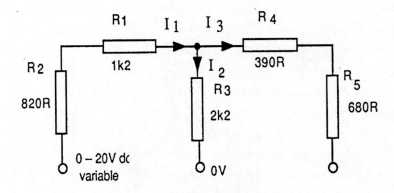

Fig 5.1

This is a similar network to the one used to investigate Kirchhoff's Laws, except that two sources of emf are used.

Monitor the variable d.c voltage.

Switch on the psu and set the variable d.c voltage to 20V.

Firstly, measure the current in each branch of the network. Do this by disconnecting each of the links 1–3 in turn, replacing them with the 0-10mA meter. This will give the current in R_1 and R_2, R_3 and R_4 and R_5 respectively.

Note both the magnitude and the polarity of each current, and tabulate them.

Question

1   Do the current directions agree with those shown in fig 5.1?

If the answer to Question 1 is no, don't alter the directions of the arrows, but mark the current as negative (see fig 5.3).

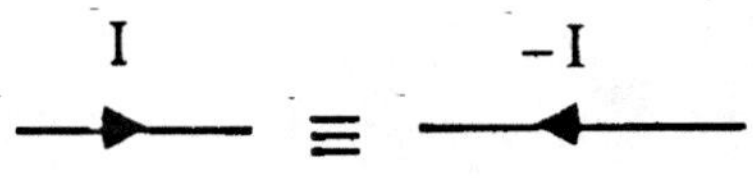

$$I \qquad\qquad -I$$

Fig 5.3

Now disconnect the 15V source and link the resistors R_3 and R_5, as shown in the circuit of fig 5.4.

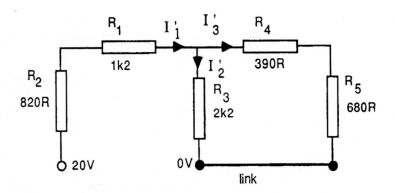

Fig 5.4

Measure and tabulate the magnitude and polarity of the currents I_1', I_2', and I_3'

Question   **2   Again, do the directions of the currents agree with those shown on the diagram?**

If they do not, show the current as negative.

Do not reverse the arrows.

Remove the link between R_3 and R_5 and replace the +15V source connections as they were initially.

Disconnect the 20V source, and link R_2 and R_3, giving the circuit shown in fig 5.5.

Copy the table as shown in fig 5.6, reproduced at the end of this assignment.

Measure the branch currents $I_1"$, $I_2"$ and $I_3"$ as before, and enter all the currents in the table.

31

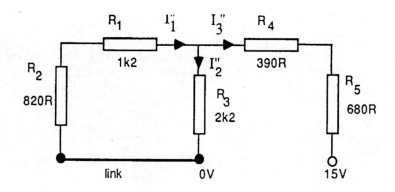

Fig 5.5

Questions

3 Can you notice any relationship between I_1, I_1' and I_1''?

4 Does the same relationship hold for I_2 with I_2' and I_2'', also I_3 with I_3' and I_3''?

You should have found that the sum of the currents due to individual voltage sources is equal to the current resulting when both sources are present in the network.

i.e $\quad I_1 = I_1' + I_1''$

similarly $\quad I_2 = I_2' + I_2''$

and $\quad I_3 = I_3' + I_3''$

This phenomenon is known as **superposition**.

If the voltage sources had any internal resistance, for example the internal resistance of a cell*, this is usually regarded as being in series with the voltage source, and thus the source would have to be replaced by its internal resistance to arrive at the correct answer. See fig 5.7.

*The internal resistance of a typical dry cell may be between 0.5 and 5 ohms and is caused by contact resistances, the resistance of the electrolyte, any corrosion of the plates, etc.

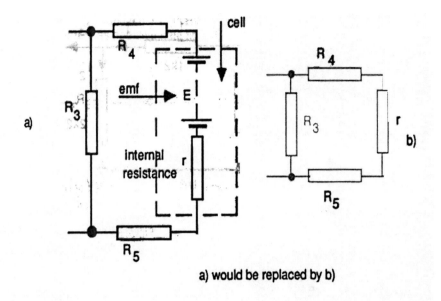

a) would be replaced by b)

Fig 5.7

The principle of superposition is summarised by the **Superposition Theorem**, which states:

'In any network containing more than one source of emf, the resultant current in any one branch is the algebraic sum of the currents that would be produced by each emf, acting alone, all the other sources of emf being replaced meanwhile by their respective internal resistances.'

Calculate, either using Kirchhoff's Laws, or Maxwell's Circulating Current Method, the currents in the total network in fig 5.1.

Questions

5   *Do your calculations agree with the experimental values found?*

Calculate the currents in the network shown in fig 5.4, and also those in the network of fig 5.5.

6   *Do these agree with experimental findings?*

7   *Does the algebraic sum of the currents due to individual sources equal the total currents due to the two sources?*

— SECTION B

THEVENIN'S THEOREM

THÉVENIN'S THEOREM

OBJECTIVE

To find a method of simplifying a network in order to obtain the current flowing in one particular branch of the network.

EQUIPMENT REQUIRED

Qty	Apparatus
1	Electricity & Electronics Constructor EEC470
1	Basic Electricity Kit EEC471
2	Multimeters or
1	Voltmeter 0–20V d.c
1	Milliammeter 0–10mA d.c
1	Power supply unit 0–20V variable d.c regulated (eg Feedback Power Supply PS445)

PREREQUISITE ASSIGNMENTS

KNOWLEDGE LEVEL

See prerequisite assignments

Lab Report Due:

Experiment Circuit

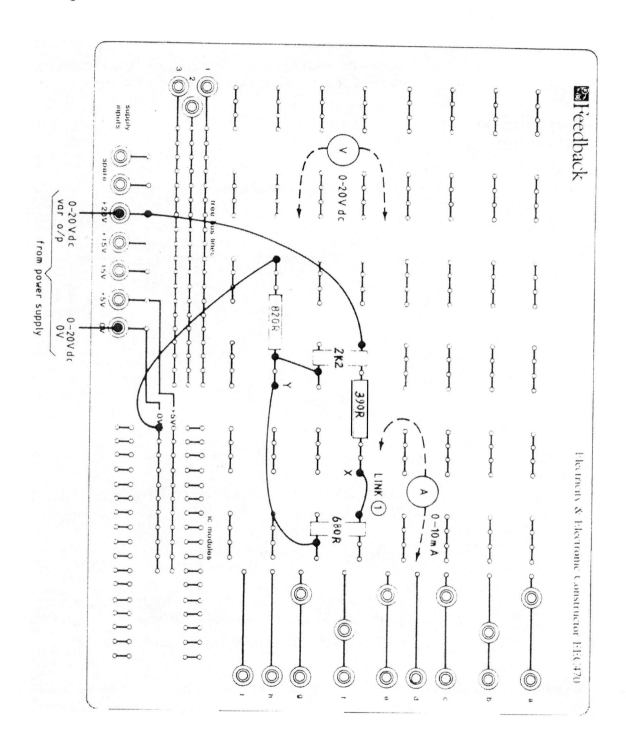

Figure 6.2

35

- Experiment Procedure

Connect the power supply unit to the mains supply line. Ensure that the variable d.c control is at minimum. DO NOT switch on yet.

For this investigation we will use the network of fig 6.1

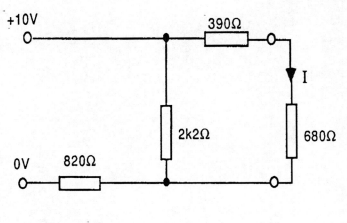

Fig 6.1

To achieve this, connect up the circuit as in the layout of fig 6.2 and check against the circuit diagram of fig 6.1 above.

Monitor the d.c input voltage.

We wish to obtain the current I, so remove link 1, and replace it with the 0-10mA meter. Switch on the psu and adjust the output voltage to 10V.

Measure and record the current flowing in the 680Ω resistor:

Question   *1   The current I, was .... milliamp?*

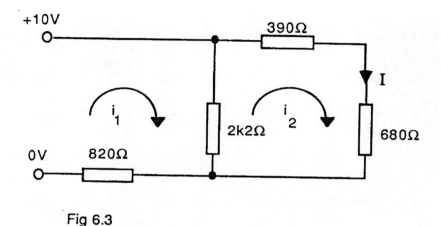

Fig 6.3

To calculate the current in the 680Ω resistor we may use Maxwell's Circulating Currents, as below:

$$10 = 3.02 i_1 - 2.2 i_2 .... \quad (1)$$
$$0 = -2.2 i_1 + 3.27 i_2. \quad (2)$$

Multiplying (2) by $\dfrac{3.02}{2.2}$

$$0 = -3.02 i_1 + 4.49 i_2 \quad (3)$$

Adding (1) and (3)

$$10 = 0 + 2.29 i_2$$
$$\therefore i_2 = 4.36 mA$$

It will be noticed that all resistance values have been worked in kΩ, thus giving an answer directly in mA.

2 Does the measured value of current correspond with the calculated value?

Although the network used is a simple one, the calculations above are fairly involved, and it can be imagined that for more complex networks the amount of mathematics needed to calculate the required branch current could be fearful. Thus a method of simplifying the network to allow more easy calculation is desirable.

The current flowing through any branch of a network is determined by the voltage across that branch, and the magnitude of the branch resistance; thus it should be possible to regard the rest of the network, other than the branch in question, as a voltage source with a series resistance. This is shown in fig 6.4.

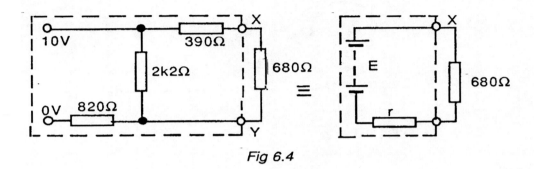

Fig 6.4

As is usual with a source, E represents the open circuit voltage of the source, and r its internal resistance.

Let us now measure the open circuit voltage of the source, and its resistance.

To measure the voltage, disconnect the 680Ω resistor from the circuit by removing the milliammeter, and measure the voltage between points X and Y as shown in fig 6.2, with the 0-20V meter.

This will give a value for E.

3   The magnitude of E was found to be .... volts?

Now we wish to determine the internal resistance r of the equivalent source. The internal resistance of a source is the resistance seen between the terminals of the source (X and Y in this case) with the sources of emf removed.

Thus remove the source of emf, E, by switching off the psu, removing the lead connecting the 'supply inputs +20V' terminal to the junction of the 2k2 and 390Ω resistors, and transferring the lead connected to the 820Ω resistor from the 'supply inputs 0V' terminal to the junction of the 2k2 and 390Ω resistors.

This gives the network shown in fig 6.5.

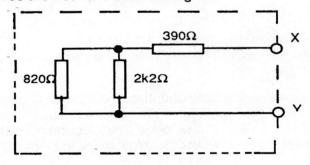

Fig 6.5

The resistance of the above network may be found by connecting a voltage to points X and Y, and measuring the resultant current. The patching diagram for this circuit is given in fig 6.6.

Switch on the psu and measure the current at voltages of 2V, 4V, 6V and 8V.

Copy the results table as shown in fig 6.7, reproduced at the end if this assignment, and enter your results.

Calculate the resistance using Ohm's Law, and take an average of the values found.

Thus, we have now found the values of E and r.

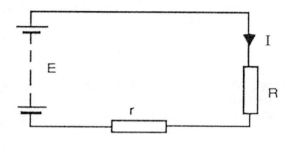

Fig 6.8

It can be seen from fig 6.8 that the required current, I, can easily be found by dividing the emf (E) by the total resistance (R + r).

i.e $$I = \frac{E}{R + r}$$

where R = resistance of required branch

Calculate $\frac{E}{R + r}$ and compare the value with the current measured initially in the 680Ω resistor

You should find that these are the same, thus we can say that: to find the current in any branch of a network, we must remove that branch and find the open-circuit voltage, then remove the sources of emf and find the resistance of the remainder of the network seen between the points at which the branch is to be connected. The current is then given by dividing the open-circuit voltage by the resistance found.

If the source of emf has any appreciable internal resistance itself, this must be added in series with the source to given the correct total resistance.

The theoretical basis for this procedure is known as **Thévenin's Theorem**, which states:

> **'The current through a resistance R connected across any two points X and Y of a network containing one or more sources of emf is obtained by dividing the p.d between X and Y, with R disconnected, by (R + r), where r is the resistance of the network measured between points X and Y with R disconnected and the sources of emf replaced by their internal resistances'.**

This theorem is often used, not in experimental work, but to simplify the calculation needed for complex networks. The calculation to solve for the current in the 680Ω resistor in the tested network, fig 6.1, is shown opposite.

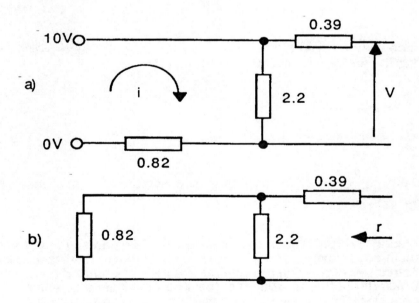

All values of resistance in kΩ

Fig 6.9

- THERETICAL RESULT

From fig 6.9 the open-circuit voltage = V

$V = 2.2\,i$

$i = \dfrac{10V}{2.2 + 0.82} = \dfrac{10}{3.02}\,mA$

$\therefore\; V = 2.2 \times \dfrac{10}{3.02} = 7.3 \text{ volts}$

$r = 0.39 + \dfrac{2.2 \times 0.82}{3.02} = 0.99k\Omega$

$\therefore R + r = 0.68 + 0.99 = 1.67k\Omega$

$\therefore I = \dfrac{V}{R + r} = \dfrac{7.3}{1.67} = 4.37mA$

Thus the current I in the 680Ω resistor can be found without resorting to simultaneous equations, as had to be done with Maxwell's Circulating Currents.

In a simple network such as the one used above the advantages of using Thévenin's Theorem may be small, but on more complex networks the advantages are great.

voltage (V)	current (mA)	resistance (Ω)
2		
4		
6		
8		
average value		

Fig 6.7

QUESTION 4: Derived the equivalent Thevennin's circuit and the current I_L, which is flowing through the resistor R_L for the following transistor circuit.

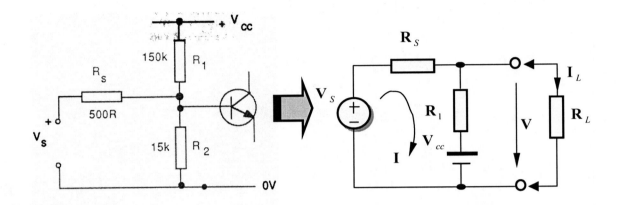

- ☐ What is open-circuit voltage V?
- ☐ What is the internal resistor r?

Hint: Using Chirchhoff's loop voltage law to get current I

$$I \times R_S + I \times R_1 + V_{cc} = V_S$$

And $\quad V = I \times R_2 + V_{cc}$

– SECTION C

POWER

To investigate the concepts of electrical power, and power transfer.

Qty	Apparatus
1	Electricity & Electronics Constructor EEC470
1	Basic Electricity Kit EEC471
1	Power supply unit 0 – 20V variable dc regulated (eg. Feedback Power Supply PS445)
2	Multimeters or
1	Voltmeter 0-20V dc
1	Milliammeter 0-20mA dc

- Experiment Procedure

Connect the power supply unit to the mains supply line. Ensure that the variable d.c control is at minimum. DO NOT switch on yet.

In the Introduction, at the beginning of this book, it was explained that work is needed to move a charge through a circuit, and that the unit of work or energy is the joule. Also the unit, the volt, was defined as the p.d between two points that exists when it takes one joule of work to move one coulomb of charge from one point to another.

Now work can be done, and energy expended at different rates. For example, take the job of moving a large mound of earth. The same amount of work and energy has to be supplied to move the earth, whether the moving is done by a man with a shovel, or by a bulldozer. However, we say that a bulldozer is much more powerful than a man and thus the job can be done far more quickly with a bulldozer. We define power as the rate of doing work. We may rate the bulldozer at, say, 100 horsepower, whereas a man may be about one half horsepower; thus for any particular sized mound of earth, a bulldozer will move it in one two-hundredth of the time a man would take.

$$.e \ Power = rate \ of \ doing \ work$$
$$= joule \ per \ second$$

or mathematically

$$Power = \frac{d}{dt} \ (work) \ where \ \frac{d}{dt} \ means \ rate \ of \ change \ with \ time$$

We said previously that:

$$One \ volt = one \ joule \ per \ coulomb$$

i.e if a charge, in coulomb, is denoted by q and work, in joule, is denoted by Wk.

$$V = \frac{Wk}{q}$$
$$.e \ Wk = qV$$
$$as \ Power = \frac{dWk}{dt}$$
$$Power = \frac{d(qV)}{dt}$$

But if the voltage is kept constant then this becomes

$$Power = \frac{Vdq}{dt}$$

or Power = Voltage x rate of change of charge

44

Now, the rate of change of charge with time is called current.

Power = Voltage x Current

i.e W = V

The unit of power is the watt. It has the dimensions of joule/second.

Now, a resistor is part of an electric circuit, and it will take work to push a charge through the resistor against the opposition offered to the charge. The power used in the resistor (we say 'dissipated by the resistor') is given by the formula above, where V is the voltage across the resistor, and I the current through it.

Let us do some measurements on the circuit of fig 7.1

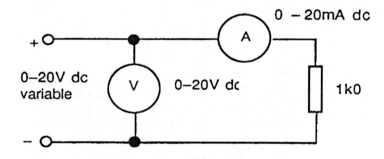

Fig 7.1

Connect up the circuit as shown in the layout of fig 7.2 opposite

Monitor the variable d.c voltage.

Switch on the power supply, and take measurements of current for voltage settings of 0V, 2V, 4V, 6V, 8V and 10V.

Copy the results table as shown in fig 7.3, reproduced at the end of this assignment, and enter your results.

45

Draw graphs as shown in fig 7.4 of voltage against current, and voltage against power, using the same axes with different scales as shown.

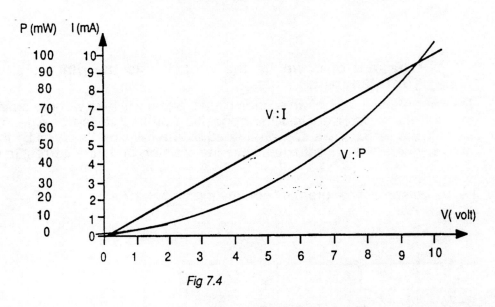

Fig 7.4

Questions

1 From your results obtained can you deduce the law of the curve for power against voltage?

2 If you double the applied voltage from 2V to 4V, how many times does the power increase?

3 If you double the applied voltage from 4V to 8V does the power increase by the same factor as it did from 2V to 4V?

4 If you triple the applied voltage, how many times does the power increase?

You should find that the power is proportional to the square of the voltage.

5 What relationship is there between the power and the current flowing?

Summarising these results mathematically we can say:

   Power $\propto$ (voltage)^2
   or Power $\propto$ (current)^2

for a resistive circuit such as fig 7.1. This can be obtained from the expression

   $W = VI$

and from Ohm's Law as below:

   $W = VI$, but $V = IR$

   $W = IR \times I \qquad \therefore W = I^2R$

   .e $W \propto I^2$

and $W = VI$, but $I = \dfrac{V}{R}$

   $\therefore W = V \times \dfrac{V}{R} \qquad W = \dfrac{V^2}{R}$

   i.e $W \propto V^2$

Summarising

   $W = V \times I$

   $W = \dfrac{V^2}{R}$

   $W = I^2 \times R$

When you have completed all the measurements calculate $I^2 R_L$

Plot a graph of Power against R_L

Mark a vertical line at $R_L = r$,

i.e $R_L = 470\Omega$

6 What is the power dissipated in R_L at $R_L = 470\Omega$?

You should find from your graph that the maximum power dissipated in the load is when the load resistance is equal to the equivalent series internal resistance of the source. In this case that is 470Ω.

We say that the load and source are matched when the conditions for maximum power transfer are met. To achieve maximum efficiency for a circuit, the source and load should be matched.

When the source and load are matched the voltage across the load is $\frac{1}{2}$ x equivalent source emf.

PRACTICAL CONSIDERATIONS AND APPLICATIONS

Power will be dissipated whenever a current flows in a resistor. This power is dissipated in the form of heat and will thus tend to heat up the resistor body.

For efficient and correct working this heat must be dispersed from the resistor body, or the temperature of the resistor will rise above permissible limits. Generally, the larger the resistor body is physically, the more power can be dissipated by the resistor without the temperature increasing too much.

Thus the resistor sizes normally used are dependent on the maximum permissible average power dissipation by the resistances in circuit. The Practical Considerations discussed at the end of Assignment 2 show the different sizes and types of resistor.

Generally, in electronic circuits, resistor power ratings are fairly low, the most common being:

$\frac{1}{10}$ W, $\frac{1}{8}$ W, $\frac{1}{4}$ W, $\frac{1}{2}$ W, 1 W, 2 W, 3 W, 5 W

perhaps the most commonly used of these are:

$\frac{1}{8}$ W, $\frac{1}{4}$ W, $\frac{1}{2}$ W resistors

48

voltage (V)	current I (mA)	power (P=VI) (mW)
2		
4		
6		
8		
10		

Fig 7.3

Load resistance (R_L) Ω	Current (I) mA	Power ($P=I^2R_L$) mW
100		
220		
390		
680		
820		
1k		

Fig 7.7

LAB 3
RESISTOR, CAPACITOR & INDUCTOR IN AC

— SECTION A

R-C CIRCUIT TIME CONSTANT

OBJECTIVE	To investigate the factors determining the charge and discharge times for a capacitor and resistor circuit..
EQUIPMENT REQUIRED	Qty Apparatus
	1 Electricity & Electronics Constructor EEC470
	1 Basic Electricity Kit EEC471
	1 Power supply unit 0 – 20V variable dc regulated. (e.g Feedback PS445)
	1 Multimeter or
	1 Voltmeter 0-20V dc
	1 2 – Channel oscilloscope (long persistence or storage type).
	1 VLF square wave generator 1Hz to 5Hz (e.g Feedback FG601)
PREREQUISITE ASSIGNMENTS	Assignment 9
KNOWLEDGE LEVEL	Before working this assignment you should :

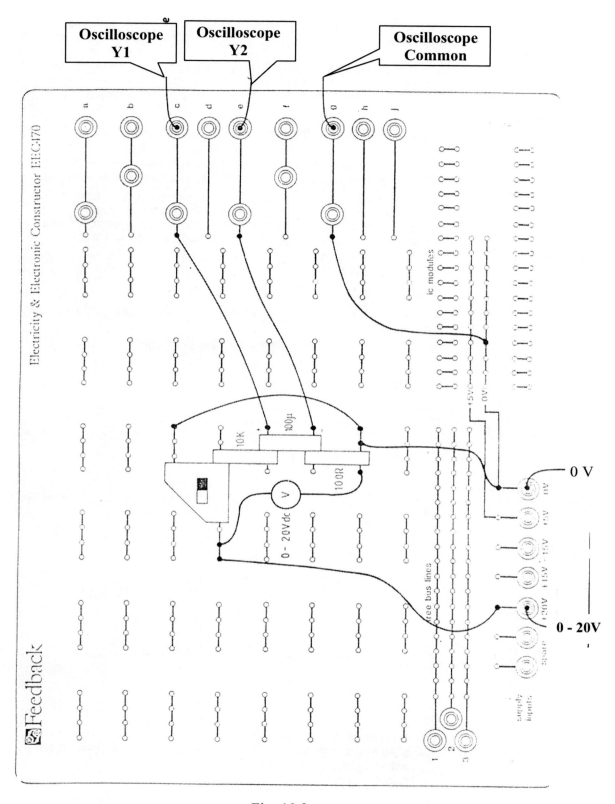

Fig. 10.2

Connect the power supply unit to the mains supply line. Ensure that the variable d.c control is at minimum. DO NOT switch on yet.

EXPERIMENTAL
PROCEDURE

We have discovered that when a capacitor is connected to a source of p.d the plates, or conductors, forming the capacitor, become charged .

We have seen that at the instant the supply is connected a current flows, the value of which is a maximum initially, and decreases with time to reach zero.

The maximum value of the current that flows we found was limited by the value of the resistor. When the current was a maximum (at time = zero) the voltage across the capacitor was zero, and the total supply voltage was being dropped across the resistor, thus giving a current determined from Ohm's Law:

$$I_{max} = \frac{V_{supply}}{R}$$

Let us now investigate further into how the capacitor charges, and what factors determine this.

Connect up the circuit as shown in the patching diagram of fig 10.2.

Set the slide switch initially to the right.

This corresponds to the circuit diagram of fig 10.1.

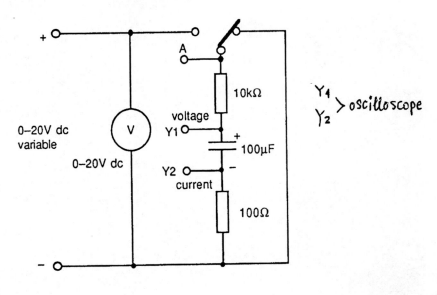

Fig 10.1

In the circuit of fig 10.1 you will notice two resistors. An oscilloscope is essentially a voltage measuring device, thus to measure the current in a circuit with an oscilloscope it is necessary to measure the voltage developed across a known resistor.

In the circuit we are using, every volt across the 100Ω resistor produces

$$\frac{1}{100} A = 10mA \text{ flowing in the circuit}$$

The 100Ω resistor is only 1% of the total resistance in the circuit, so it will not effect the charging greatly.

Also, as the maximum charging current is:

$$\frac{10V}{10k\Omega} = 1mA$$

then the maximum voltage across the resistor will be:

$$1mA \times 100\Omega = 0.1V$$

therefore there will only be a small error in the voltage readings as well.

So the inclusion of the 100Ω allows the current waveform to be displayed on the oscilloscope without affecting greatly the action of the circuit.

Set the sensitivity of the Y_1 channel of the scope to 2V/cm.
Set the Y_2 sensitivity to 50mV/cm.

Set the timebase to 500mS or to the slowest rate available on your instrument.
Zero both traces.
Set both the Y amplifiers to be d.c coupled.

Ensure that the variable d.c control knob is at min. Monitor the variable d.c voltage.

Switch on the power supply unit and set the variable d.c voltage to 10V.

Switch the slide switch to the left and observe the current and voltage charging waveforms on the oscilloscope.

1. **Do these waveforms correspond in shape to those plotted before?**

Switch the slide switch to the right and observe the discharge current and voltage waveforms.

2   **What shape relationship do the discharge curves bear to the charging curves?**

Draw the charge and discharge curves.

Now disconnect Y1 from the voltage monitoring point and connect it to point A

The oscilloscope is now looking at the voltage applied to the series capacitor–resistor (CR) circuit.

3   **What voltage do you measure at this point in the circuit ?.**

Operate the switch left and right regularly and observe the Y1 waveform
Draw what you see on the oscilloscope.
The waveform you observe should be of the form of fig 10.3.

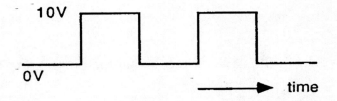

Fig 10.3

The voltage is seen to alternate between the levels of 0V and 10V

If you are switching regularly the time (ie. t)for which the voltage is at 10V is the same as the time for which it is at 0V. (See fig 10.4).

Also the time interval between successive points of repetition of the waveform (e.g A to A, or B to B) will be equal (i.e T).

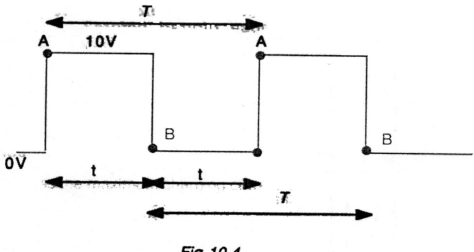

Fig 10.4

For any waveform which repeats itself regularly with time, as the one in fig 10.4 does, the time T will be constant, and is called the period of the waveform, being the period of time it takes for the wave to repeat itself.

The wave is said to make one cycle during T seconds.

In electronic work voltages and currents may alternate between levels in many ways, and extremely quickly. They may alternate many times each second.

The number of alternations made in one second is called the frequency of alternation, or just the frequency of the waveform.

The frequency of a waveform is expressed in either 'cycles per second' or 'Hertz'.

1 cycle per second (c/s) = 1Hertz (Hz)

Cycles per second is a very descriptive term, but the correct unit to use is Hertz, named after Heinrich Hertz, a German scientist who worked on electrical experimentation.

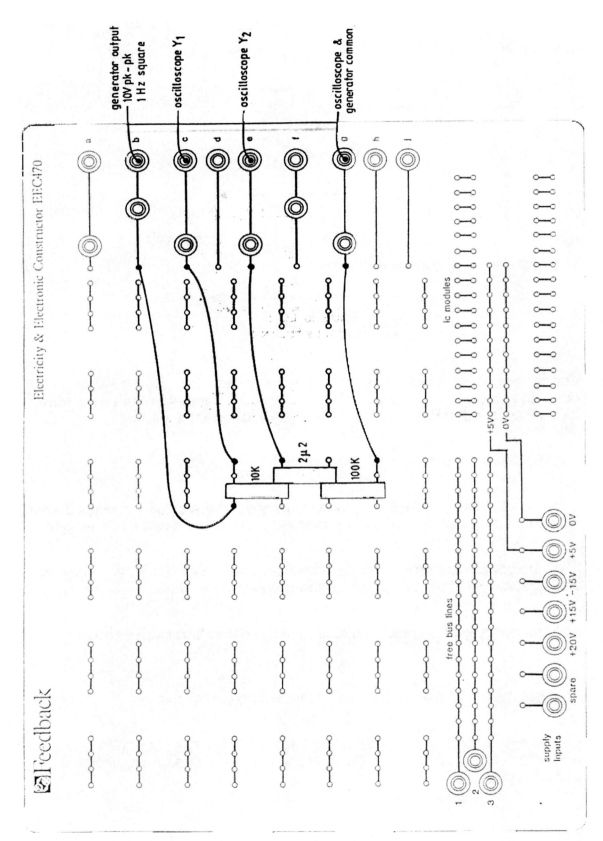

Fig 10.6

56

Connect the circuit as shown in the patching diagram of fig 10.6 corresponding to the circuit diagram of fig 10.5.

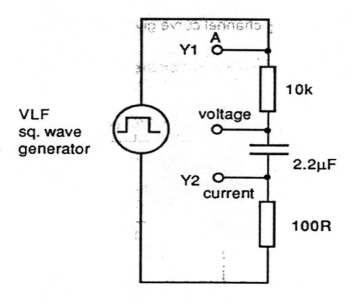

Fig 10.5

Connect the oscilloscope Y_1 input initially to point A.

Set the generator so that the output square wave from it has a frequency of, say, 1Hz, at an output level of 10V pk–pk.

4   Is the waveform the same shape as fig 10.3 ?

5   What is the period of the waveform?

Set the frequency to 2Hz.

6   What is the period of the waveform now?

Mark out a piece of graph paper with axes as in fig 10.7 and draw the input waveform on the top axis.

Disconnect the Y_1 input from point A, and reconnect it to monitor the voltage across the capacitor.

7   Is the voltage curve of the same shape as expected?

The Y1 curve shows the voltage charge and discharge curves one after the other.

The Y$_2$ channel curve gives the current charge and discharge.

Draw these curves on the axes indicated in fig 10.7.

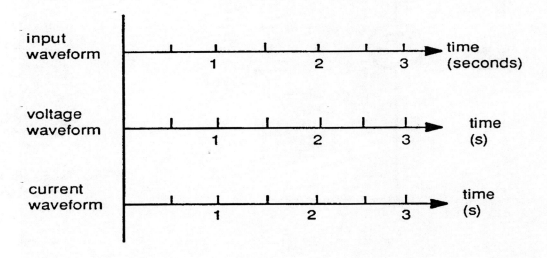

Fig 10.7

The type of curves that you have drawn is called *exponential*. This is the mathematical name for such curves, as they follow an exponential law.

This law for the voltage is:

$$V_C = V (1 - e^{-t/CR}) \qquad (1)$$

where

    V_c = voltage across the capacitor

    V   = supply voltage

    t   = time (second)

    R   = resistance (ohm)

    C   = capacitance (farad)

    e   = exponential constant = 2.718

The law for the current is:

$$i = \frac{V}{R} e^{-t/CR} \qquad (2)$$

where i   = current in circuit.

Equations (1) and (2) obviously define how the voltage and current behave for any particular R and C. From (1) it can be seen that, as t increases, V_c becomes nearer and nearer to V but will not reach it until $e^{-t/CR}$ becomes zero. For $e^{-t/CR}$ to become zero, t must before infinite. Thus the capacitor will never become fully charged.

Similarly, from equation (2), as t increases i becomes smaller and smaller, but will only reach zero when t reaches infinity.

From equations (1) and (2) it can be seen that the rate of charge is dependent on the product CR.

At t = 0 the capacitor potential is 0 and at t = ∞ it is V. Between these limits the change in voltage V_c will occur according to the exponential curves, dependent only on CR.

Consider equation (2)

$$i = \frac{V}{R}(e^{-t/CR})$$

at t = 0 the initial current (rate of charge) is given by

$$i = \frac{V}{R}(e^{-0/CR})$$

If this current was maintained constant throughout charging of the capacitor, the time taken to reach the fully charged state would be, say, T seconds. At this time the charge on the capacitor would be Q coulombs where:

$$Q = CV$$

now also

$$Q = iT$$

$$\therefore CV = \frac{V}{R}.T$$

$$\therefore T = \frac{CV}{\frac{V}{R}}$$

$$\therefore T = CR \text{ seconds}$$

This time is called the TIME CONSTANT (γ)
of the resistor-capacitor circuit.

This is shown in fig 10.8.

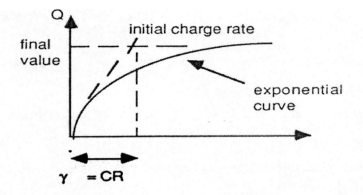

Fig 10.8

Clearly the time taken for the capacitor to charge will be longer than if the current had remained constant. The actual value that the voltage reaches at $t = \gamma$ is found from equation (1).

$$V_C = V(1-e^{-\gamma CR})$$
$$V_C = V(1-e^{-1})$$

From which

$$V_C = V(0.632)$$

Thus V_C will reach 63.2% of the supply voltage in the time γ

Calculate the time constant of your circuit.

Calculate 63.2% of your input voltage.

Measure the time taken for the voltage waveform to reach 63.2% of the input voltage.
Compare this with the calculated time-constant.

8   What was the initial value of current?

9   What value of current was reached in time γ ?

1 0   What percentage of the initial value of current is this?

60

From equation (2), substitute $t = \gamma = CR$, as below, and calculate i as a percentage of the initial current, $\dfrac{V}{R}$

$$i = \frac{V}{R}(e^{-1})$$

$$\therefore \ i = 36.8\,\%\,\frac{V}{R}$$

1 1   *Does this value agree with your measured value?*

PRACTICAL CONSIDERATIONS AND APPLICATIONS

We have examined equation (1) and deduced that during the time from $t = 0$ to $t = \gamma = CR$ the voltage on the capacitor will rise to 63.2% of its final value, the supply voltage V. The difference between V_c and V will thus be 36.8%V at this point in time. This difference is sometimes called the voltage margin.

It can be shown, from equation (1), that during each period of γ the voltage across the capacitor increases by 63.2% of the voltage margin at the beginning of that period. This is shown in fig 10.9.

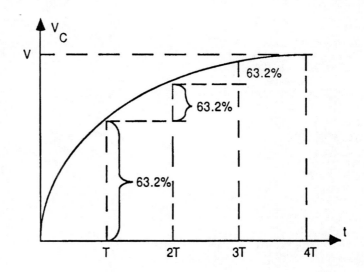

Fig 10.9

61

— SECTION B

RMS VALUE OF AN AC WAVEFORM

OBJECTIVE

To investigate a.c waveforms and the power carried by them.

EQUIPMENT REQUIRED

Qty	Apparatus
1	Electricity & Electronics Constructor EEC470
1	Basic Electricity Kit EEC471
1	Power supply unit 0 – 20V variable dc regulated and 5V a.c (eg. Feedback Power Supply PS445)
2	Multimeters or
1	Voltmeter 0-20V dc and
1	Milliammeter 0-100mA dc
1	2–Channel oscilloscope

PREREQUISITE ASSIGNMENTS

KNOWLEDGE LEVEL

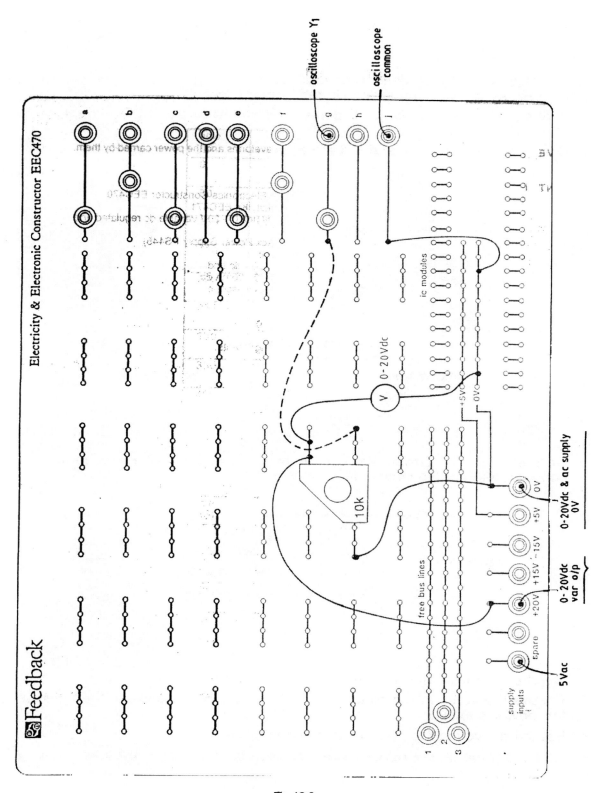

Fig 13.3

Connect the power supply unit to the mains supply line. DO NOT switch on yet.

In Assignment 10 we defined several terms relating to varying waveforms. Let us now re-examine these ideas in rather more depth.

A direct current or voltage is generally thought of as unidirectional and not varying with time.

Graphs of direct current and voltage against time are of the form of fig 13.1.

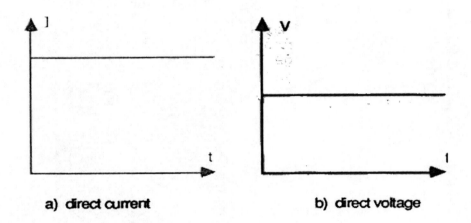

a) direct current

b) direct voltage

Fig 13.1

Connect up the circuit as shown in the patching diagram of fig 13.3 corresponding to the circuit diagram of fig 13.2.

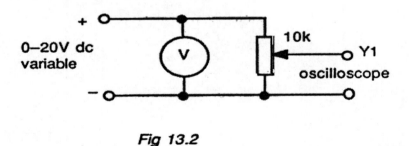

Fig 13.2

Set the potentiometer to its mid position.

Ensure the variable d.c control knob is fully counterclockwise, then switch on the power supply,.

Set the d.c voltage to 10V, as shown on the meter.

Zero the trace on the oscilloscope and then set the Y sensitivity to 5V/cm.

Observe the trace. You should get a graph such as fig 13.1(b) with the voltage level around 5V.

This trace shows a direct voltage.

Now vary the setting of the potentiometer about its mid position.

1.   *As you vary the setting of the potentiometer what does the voltage, as seen on the oscilloscope, do?*

2 .   *Does it ever cross the zero voltage axis?*

A voltage that is always of the same polarity (i.e one that does not pass through the zero voltage axis), but varies with time may be called a varying direct voltage. Fig 13.4 shows such a voltage.

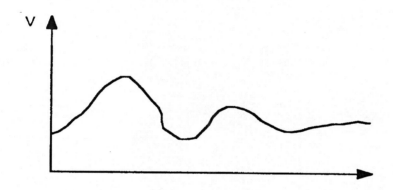

Fig 13.4

It is possible for a voltage to vary so that a reversal of polarity occurs. In such cases the term *alternating* voltage is used.

Transfer the oscilloscope connection to the 5V a.c supply.

Switch the timebase to 5ms/cm.

Draw the waveform you see.

3 .   *What is the time period between two successive points of repetition of the waveform?*

4 .   *The period of the waveform is ......milliseconds?*

5 .   *If the period of the waveform is as found above, how many complete variations does the voltage make each second?*

The number of complete variations, known as a 'cycle' made each second is called the FREQUENCY of the waveform. Frequency is measured in cycles per second, or more correctly Hertz.

1 cycle/second (c/s) = 1 Hertz (Hz)

6. *The frequency of the waveform examined was ......Hz?*

7. *What is the relationship between the period and the frequency of a waveform?*

Measure the peak voltage of the waveform.(refer to fig 13.5

8. *The peak-to-peak voltage was .... volt?*

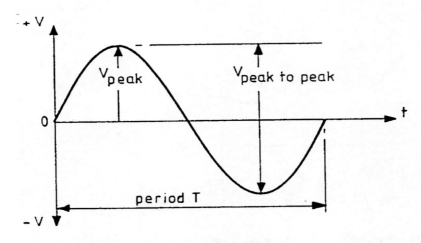

Fig 13.5

9. *The peak voltage was .... volts?*

10. *What is the relationship between the peak and peak-to-peak voltages?*

You should find that the period and the frequency are reciprocals of one another:

$$T = \frac{1}{f} \quad \text{and} \quad f = \frac{1}{T}$$

and that $V_{pk\text{-}to\text{-}pk} = 2 \times V_{pk}$.

Let us now investigate the power producing properties of some waveforms.

First of all let us consider the direct current waveform of fig 13.6, $i = I$ (constant)

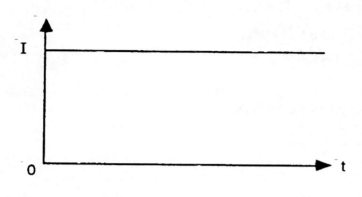

Fig 13.6

Suppose the current is passed through a resistor of value R ohms

At all times the current value is I A and the resistor value is R ohms, thus the power dissipated in the resistor will be:

Power = I^2 R watts

Suppose now that a current with the waveform shown in fig 13.7 is passed through the resistor R.

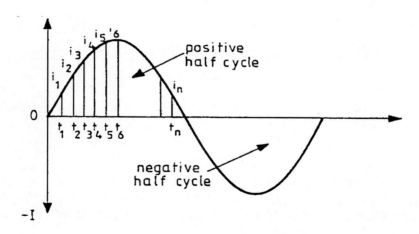

Fig 13.7

At time t_1 the power will be $i_1{}^2 R$ watts

At time t_2 the power will be $i_2{}^2 R$ watts

At time t_3 the power will be $i_3{}^2 R$ watts

At time t_n the power will be $i_n{}^2 R$ watts

The average power produced will be:

$$= \frac{i_1{}^2 R + i_2{}^2 R + i_3{}^2 R + \ldots i_n{}^2 R}{n} \text{ watts}$$

This is the average power produced in the positive half-cycle shown in fig 13.7. The power produced in the negative half-cycle will be the same because the currents, although negative, will have positive squares,

i.e $\quad (-i_1)^2 R = i_1{}^2 R$

Thus there will be an effective power value of current for the a.c waveform. The effective value I is that value of direct current which produces the same power as the average power produced by the a.c.

i.e where $I^2 R = \dfrac{i_1{}^2 R + i_2{}^2 R + i_3{}^2 R + \ldots i_n{}^2 R}{n}$

$$I^2 = \frac{i_1{}^2 + i_2{}^2 + i_3{}^2 + \ldots i_n{}^2}{n}$$

$$\overline{I^2} = \sqrt{\frac{i_1{}^2 + i_2{}^2 + i_3{}^2 + \ldots i_n{}^2}{n}}$$

The effective value of current is found from the root of the mean of the squares of the currents in the a.c waveform, and is thus termed the ROOT MEAN SQUARE value, or the rms value of the a.c waveform.

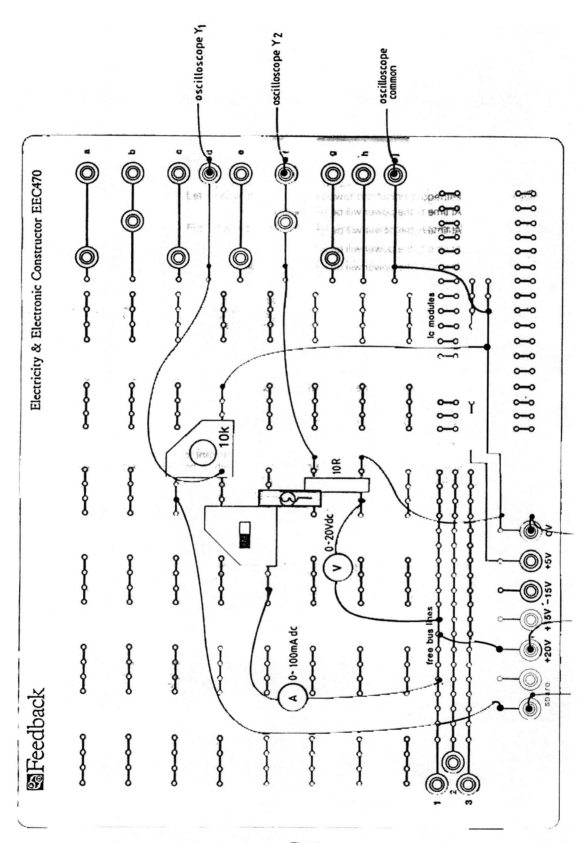

Fig 13.9

Ensure that the variable d.c voltage control is fully counterclockwise, and switch off the psu .

Connect up the circuit as shown in the patching diagram of fig 13.9, corresponding to the circuit diagram of fig 13.8.

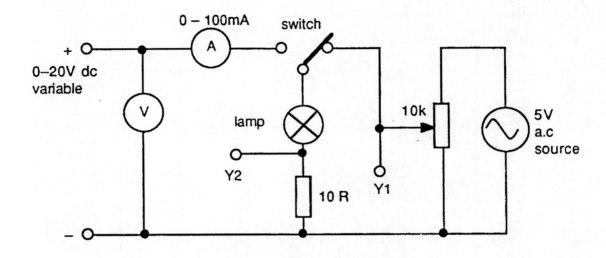

Fig 13.8

Set the potentiometer to its mid position and switch on the psu

Set the slide switch initially to the left

Set the variable d.c voltage to about 5V as shown on the meter. The lamp should be on, but dim. Notice the intensity of the lamp.

Switch the slide switch to the right and adjust the potentiometer untill the intensity of the lamp is roughly the same as before.

Switch the slide switch left and right, adjusting the potentiometer at the same time, until the intensity of the lamp is identical for both switch positions.

Copy the results table as shown in fig 13.10, reproduced at the end of this assignment, and enter your results.

Take readings of the d.c voltage and current, and the a.c peak-to-peak voltages seen on the scope.

Calculate the peak-to-peak a.c current.

11. As the lamp is at the same intensity for the a.c and the d.c waveforms, what can you say about the power in the a.c and d.c waveforms?

12. What is the effective value of the alternating current?

13. What is the rms value of the alternating voltage?

14. What is the peak value of the alternating voltage?

15. What is the peak value of the alternating current?

Calculate the relationships between the rms values and the peak values for current and voltage.

The ratio: $\dfrac{\text{peak value}}{\text{rms value}}$ is called the *Peak Factor* of the waveform.

16 What is the peak factor of the current waveform?

17 What is the peak factor of the voltage waveform?

The waveforms you have been examining are approximately of the shape called sinusoidal. This is because the amplitude or value of the waveform at any time is related to the sine of that time. It is possible to construct a sine wave, as shown in fig 13.11 (shown opposite) as follows:

Draw a circle with a conveniently sized radius (say one inch). Mark off radii in the circle every 30°. To the right of the circle mark off a vertical axis and a horizontal scale in degrees.

From the points where the radii meet the circle, project horizontal lines to meet the vertical lines drawn from corresponding points on the horizontal scale.

The resultant waveform when these points are joined is a sine wave.

If the rotating vector OA rotates at an angular speed of ω radians per second then, as there are 2π radians per revolution, the vector will make

$\dfrac{\omega}{2\pi}$ revolutions per second.

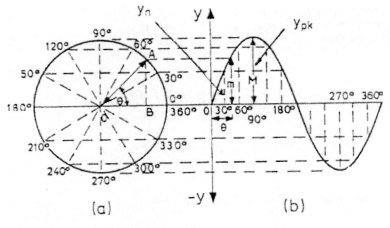

construction of a sine wave

Fig 13.11

Each complete revolution traces out one cycle of sine wave, therefore there will be:

$$\frac{\omega}{2\pi} \text{ cycles per second}$$

Thus the sine wave traced out will have a frequency of

$$f = \frac{\omega}{2\pi} \text{ Hz}$$

$$\underline{\omega = 2\pi f}$$

Draw such a sine wave, and mark off the horizontal scale in 20° intervals, i.e 20°, 40°, 60°, 80°, ... 160°, 180° to meet your waveform.

Draw vertical lines at the mid-ordinates of these intervals, i.e at 10°, 30°, 50°, ... 150°, 170°.

Calling the peak value of the sine wave unity (i.e the radius of the circle r = 1) graduate the vertical axis between -1 and +1.

Copy the table as in fig 13.12, reproduced at the end of this assignment, and enter the values of the mid-ordinates for the angles given in the table.

Calculate:

$$y^2_{10} + y^2_{30} + y^2_{50} + y^2_{70} + \ldots \cdot y^2_{170}$$

In the above, the number of mid-ordinates is 9.

$\therefore$ Calculate

$$\frac{y^2_{10} + y^2_{30} + y^2_{50} + y^2_{70} + \ldots y^2_{170}}{9}$$

Square root this to find:

$$\sqrt{\frac{y^2_{10} + y^2_{30} + y^2_{50} + y^2_{70} + \ldots y^2_{170}}{9}}$$

This will give you the rms value of a sine wave whose peak value is unity

18. *How does this compare with the peak factor that you calculated earlier?*

Multiply the rms value above by the peak value of the alternating voltage used before.

19. *How does this value compare with the rms value found from the lamp experiment?*

Do the same for the alternating current.

You should have found that the peak factor for a pure sine wave is given by:

Peak factor $= \sqrt{2}$

$= 1.414$

Also the rms value of a sine wave with unit peak value should be:

$$rms = \frac{1}{\sqrt{2}}$$

Thus we can say:

For a sine wave Vrms = 0.707V pk

If a source of triangle and/or square waves is available, repeat the experiment to find the rms values of those waveforms. Construct graphically triangle and square waves and verify your experimental results.

73

d.c voltage (V)	a.c voltage p–p (V)

d.c current (mA)	a.c current p–p (mA)

Fig 13.10

angle	y_n	y_n^2
10		
30		
50		
70		
90	1.0	
110		
130		
150		
170		

Fig 13.12

74

— SECTION C

R, L, C CIRCUITS AT AC

❑ <u>UNIT 1</u> RESISTIVE CIRCUIT AT AC

OBJECTIVE	To investigate a resistive circuit at a.c.

EQUIPMENT REQUIRED

Qty	Apparatus
1	Electricity & Electronics Constructor EEC470
1	Basic Electricity Kit EEC471
1	Power supply unit 0 – 20V variable dc regulated and 5V a.c (eg. Feedback PS445)
2	Multimeters or
1	Voltmeter 0-20V dc and
1	Milliammeter 0-50mA dc and
1	Voltmeter 0–10V ac and
1	Milliammeter 0 – 100mA ac
1	Function generator 50 – 1kHz Sine 20V pk–pk (eg. Feedback FG 601)

PREREQUISITE ASSIGNMENTS

KNOWLEDGE LEVEL

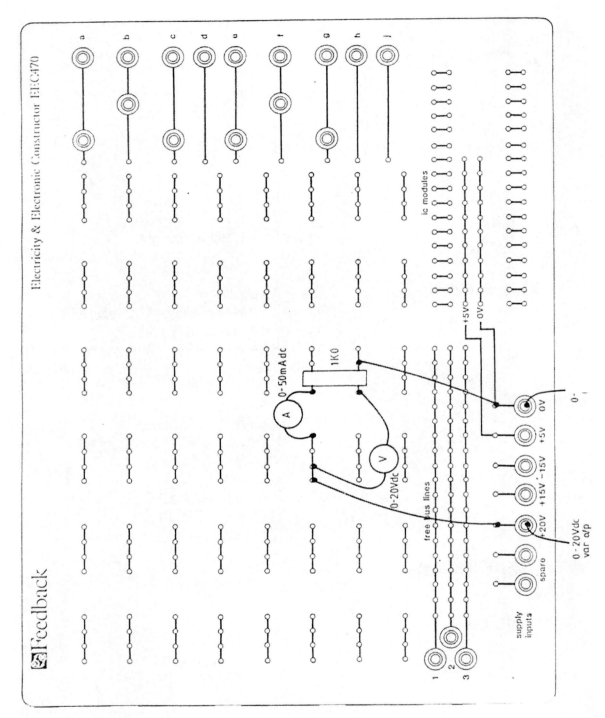

Fig 14.2

Connect the power supply unit to the mains supply line. DO NOT switch on yet.

We wish now to see how a resistor behaves at a.c.

Connect up the circuit as shown in the patching diagram of fig 14.2, corresponding to the circuit diagram of fig 14.1.

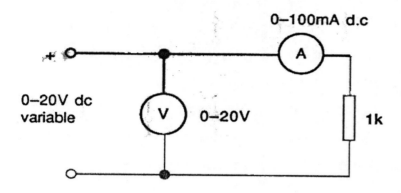

Fig 14.1

Ensure the variable d.c voltage control knob is turned fully counterclockwise, then switch on the psu.

Vary the d.c control slowly, and observe the two meters.

1. Do they keep in step with each other?

Vary the control somewhat faster.

2. Do they still keep in step with each other?

You should find that when the current is zero the voltage is zero; and, by Ohm's Law, the current is proportional to the voltage, so that if a.c is applied to a resistor the current waveform and the voltage waveform will be of the same shape. See fig 14.3.(shown overleaf)

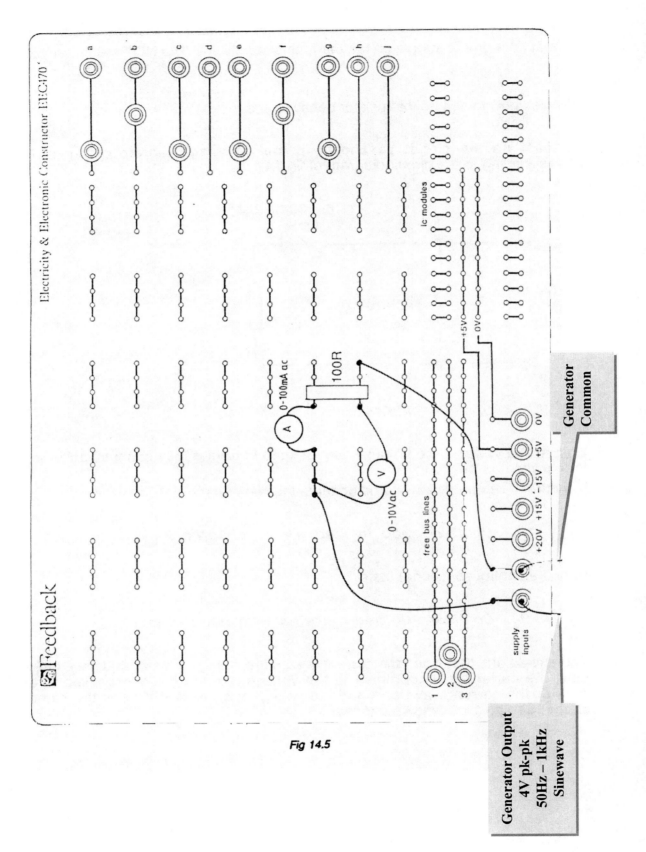

Fig 14.5

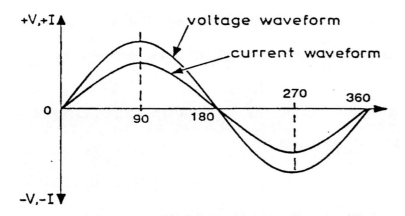

Fig 14.3

The two waveforms will have their zeros at the same time, and their maximum values at the same time.

In a resistor, the current and voltage are said to be *in phase*.

Now connect up the circuit of fig 14.4 as shown the patching diagram of fig 14.5.

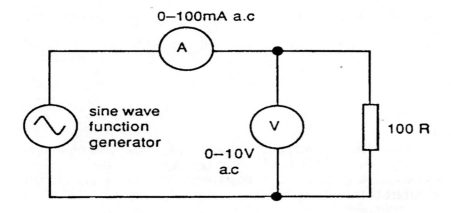

Fig 14.4

Set the generator frequency to 50Hz, with an output amplitude of 4V rms, as read on the 0-10V a.c meter.

Copy the results table as shown in fig 14.6, reproduced at the end of this assignment, and record the frequency, current, and voltage

Calculate the resistance of the resistor at 50Hz and tabulate this also.

Repeat the readings for a frequency of 100Hz, and for 100Hz increments thereafter, up to 1kHz.

3.   *How do the current, voltage and resistance change with frequency?*

It should be seen from this experiment that the current and voltage in a resistive circuit are in phase, and that the resistance of a resistive circuit does not change with frequency (see fig 14.7).

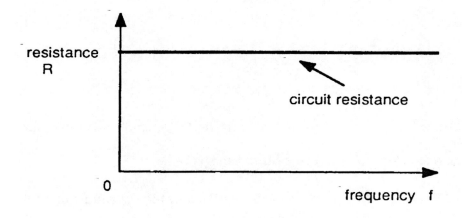

Fig 14.7

Resistors are used at a.c in similar ways to those d.c applications of earlier assignments. There is no phase shift across a true resistance, and the ratio of voltage to current shown by a resistance is constant with frequency, so the behaviour at a.c is no different from its behaviour at d.c.

When calculating power dissipation at a.c, rms values of voltage and current must apply.

frequency (Hz)	voltage (V) rms	current (mA) rms	resistance (ohms)
50			
100			
200			
300			
400			
500			
600			
700			
800			
900			
1000			

Fig 14.6

❑ UNIT 2 CAPACITIVE CIRCUIT AT AC

OBJECTIVE

To investigate a capacitive circuit at a.c.

EQUIPMENT REQUIRED

Qty	Apparatus
1	Electricity & Electronics Constructor EEC470
1	Basic Electricity Kit EEC471
1.	2–channel oscilloscope
1.	Function generator 250Hz sine 10V pk–pk (eg. Feedback FG 601)

PREREQUISITE ASSIGNMENTS

KNOWLEDGE LEVEL

Before working this assignment you should :

● Know how to describe varying ac quantities by vectors.

See also prerequisite assignments.

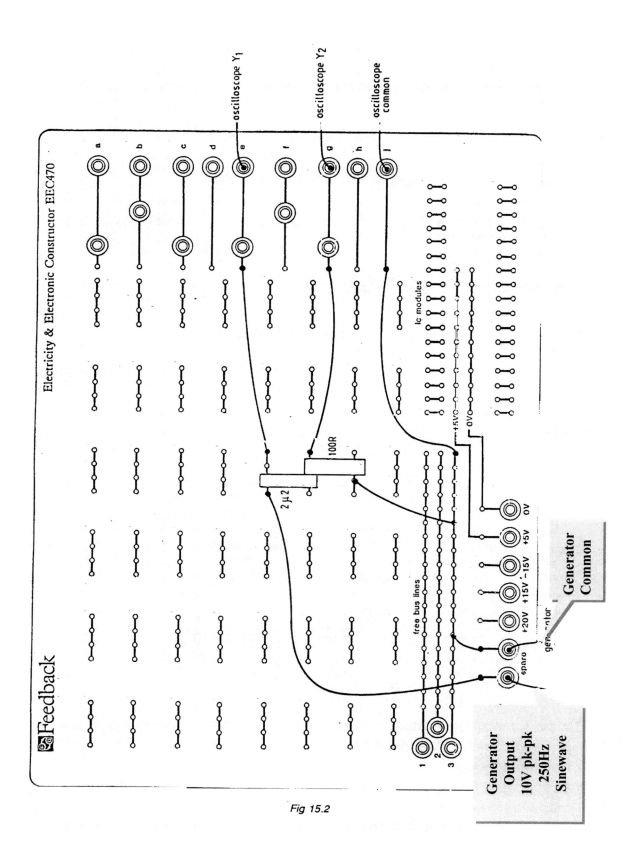

Fig 15.2

We know, from Assignment 8, that the relationship between charge, voltage and capacitance is:

$$Q = CV$$

Also we know: $Q = I\,t$, where I is current and t is time

From these we can say that if a capacitor of C farad is charged from 0V to V volts, in t seconds then:

charging current, $I = \dfrac{Q \text{ coulombs}}{t \text{ seconds}}$

charging current = capacitance x rate of increase of voltage

Let us see what happens when a sinusoidal alternating voltage is applied to a capacitor.

Connect up the circuit as shown in the patching diagram of fig 15.2 corresponding to the circuit diagram of fig 15.1.

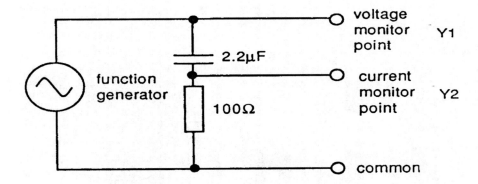

Fig 15.1

Set the function generator to give a 10V peak-to-peak sine waveform at 250Hz
Set the oscilloscope as follows:
Y1 channel 1V/cm.
Y2 channel 500mV/cm.
Timebase to 1ms/cm
Zero both the traces and then observe the two waveforms on the oscilloscope.

Carefully draw the two waveforms, showing their relative positions with respect to each other.

Look at the *voltage* waveform.

1. *Where on the waveform is the rate of change of voltage a positive maximum?*

2. *Where on the waveform is the rate of change of voltage zero?*

3 *Where on the waveform is the rate of change of voltage a negative maximum?*

We have said that the current is proportional to the rate of change of voltage.

4. *Does your current curve substantiate this?*

5. *Using the same notation as in fig 13.11 (i.e 360° in one cycle), how many degrees are the two waveforms apart?*

6. *What proportion of a cycle is that?*

7. *Remembering that the time increases towards the right, which of the two waveforms reaches its positive maximum first?*

As the current waveform reaches it positive peak value 90° before the voltage waveform reaches that value, we say that:

The current in a capacitive circuit leads the voltage by 90°.

Often the 90° is referred to as $\frac{\pi}{2}$

As there are 2π radians in 360°, then 90° is $\frac{\pi}{2}$ radians

85

Mathematically, if the voltage waveform is denoted by the formula:

$$v = V_{max} \sin \omega t$$

then as i = C x rate of change of voltage

$$i = C\frac{dv}{dt} = CV_{max}\frac{d}{dt}(\sin \omega t)$$

$$\therefore i = CV_{max}\, \omega \cos \omega t$$

Thus if v is sinusoidal, i will be cosinusoidal, i.e the same shape, but leading by 90°.

This is because $\cos \omega t = \sin(\omega t + 90°)$

Figure 13.11 shows how a sine wave can be plotted by taking projections from a rotating vector.

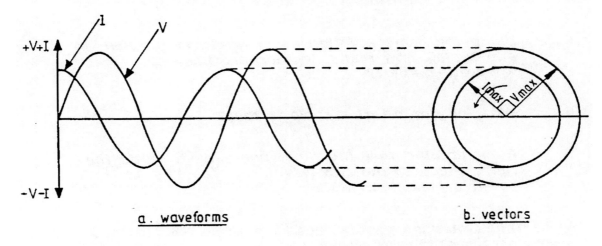

a. waveforms   b. vectors

FIG 15.3

If we were to plot the voltage and current waveforms in the capacitor by that method we would require two vectors. Both vectors would rotate while keeping a constant 90° angle between them, as shown in fig 15.3b.

As we go on we shall find it useful to think in terms of these vectors, but rather confusing if they are always rotating. Usually it is the relationships between them that are important, as for instance the 90° angle between those in fig 15.3b.

These relationships can be studied conveniently in a diagram where the vectors are shown at rest. The vectors are then said to be represented by 'phasors' and the diagram is a 'phasor diagram'. Fig 15.4 is a phasor diagram corresponding to fig 15.3b.

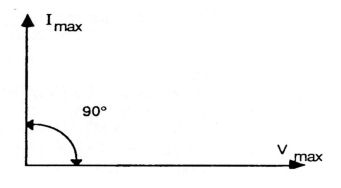

Fig 15.4

The voltage phasor is taken as the reference and is drawn horizontally pointing to the right (3 o'clock). The current phasor leads the voltage phasor by 90° and is thus drawn 90° counterclockwise from the reference.

When an alternating voltage is applied across a capacitor an alternating current flows. Yet when a d.c voltage is applied, after an initial flow of charging current, no d.c current flows. This behaviour is different from that of a resistance. Nevertheless if an a.c voltage and an a.c current can exist, the ratio between them is likely to be of interest, and the ratio is therefore given a different name.

In an a.c circuit the ratio of voltage to current is called 'impedance' and is denoted by Z.

Thus in an a.c circuit $Z = \dfrac{V}{I}$

We shall examine this idea further in other assignments. For the moment it may be noted that impedance may be taken as the ratio of two phasors, and therefore has both magnitude and phase.

$$\text{Magnitude} = \left| Z \right| = \frac{V_{max}}{I_{max}}$$

Phase of the impedance is the angle between the phasors. The impedance of a capacitor has a phase of $-90°$ or $-\dfrac{\pi}{2}$ radians

Impedances of ±90° phase angle have special properties and are given the special name '*reactance*'.

❑ <u>UNIT 3</u> INDUCTIVE CIRCUIT AT AC

OBJECTIVE

To investigate an inductive circuit at a.c.

EQUIPMENT REQUIRED

Qty	Apparatus
1	Electricity & Electronics Constructor EEC470
1	Basic Electricity Kit EEC471
1.	2–channel oscilloscope
1.	Function generator 250 Hz Sine 10V pk–pk (eg. Feedback FG 601)

PREREQUISITE ASSIGNMENTS

KNOWLEDGE LEVEL

Before working this assignment you should

● Know how to describe varying ac quantities by phasors.

● Know how a capacitive circuit behaves at ac.

● Know how to apply Kirchoff's Laws.

See also prerequisite assignments.

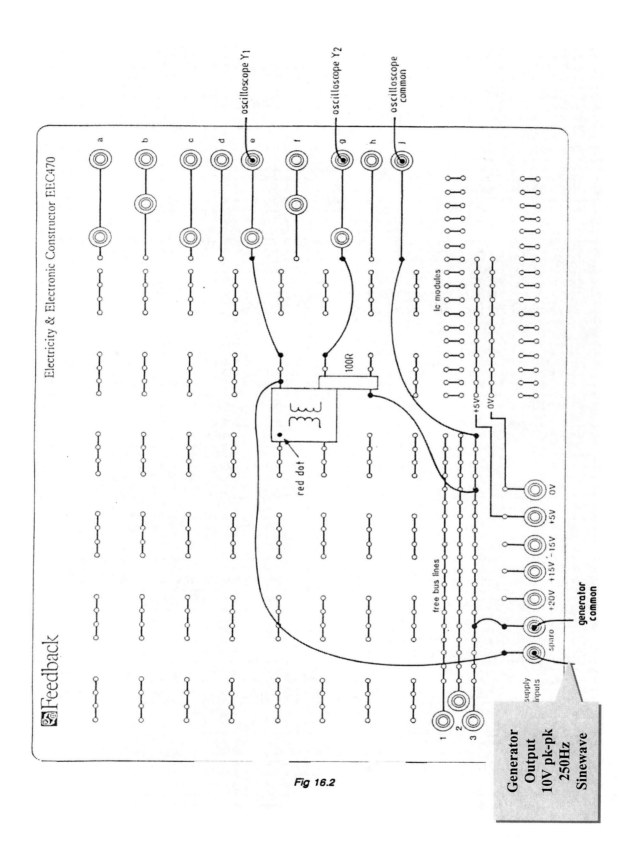

Fig 16.2

From Assignment 12 we have found out that the relationship between induced emf, current and inductance in a system is:

$$e = -L \frac{di}{dt}$$

or: (Induced emf) = − (inductance)(rate of change of current)

We will now examine what happens when a sinusoidal alternating voltage is applied to an inductor.

Connect up the circuit as shown in the patching diagram of fig 16.2 corresponding to the circuit diagram of fig 16.1.

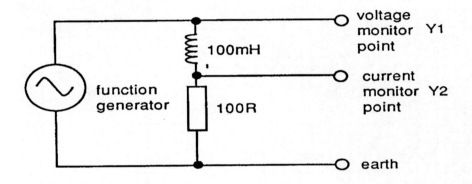

Fig 16.1

Set the function generator to give a 10V peak-to-peak waveform at 250Hz.
Set the oscilloscope as follows:
Y1 channel (current) to 1V/cm
Y2 channel (voltage) to 500mV/cm
Timebase to1ms/cm.
Zero both traces, then observe the two waveforms on the oscilloscope.
Draw the two waveforms you see, showing their relative positions with respect to each other.

1.   *Where on the waveform is the rate of change of current a positive maximum?*

2.   *Where is it zero?*

3. *Where is it a negative maximum?*

4. *Does the voltage waveform correspond to the rate of change of current?*

5. *How far apart, in degrees, are the voltage and current waveforms?*

6. *How far apart are they in radians?*

7. *Which waveform is leading?*

8. *Is this the same as for a capacitive circuit.*

If the current waveform is denoted by

$$i = I_{max} \sin \omega t$$

$$\text{then as } e = -L\frac{di}{dt}$$

$$\text{then } e = -LI_{max} \omega \cos \omega t$$

Here we must be careful. We have found the induced emf, and included the minus sign as a reminder that it opposes the change of current. But when we look at a.c applied to a resistor or d.c applied to it , we look at the *applied* voltage.

In the same way, with the inductance, we must look at the applied voltage if we are to be consistent. Since ΣV around the circuit is zero (Kirchoff's Law) , this applied emf is equal and opposite to the induced emf, i.e applied emf $= +L\frac{di}{dt}$

$$LI_{max} \omega \cos \omega t$$

So we can say that a positive applied emf produces a positive current which induces an opposing (negative) emf in the inductor.

This was the voltage you were looking at experimentally, as should be evident by a look at fig 16.3, in which the polarity of the emfs is indicated by arrows in the direction of action.

91

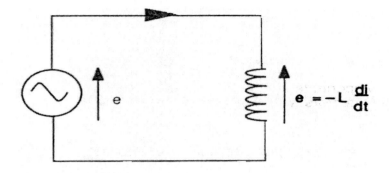

Fig 16.3

The results show that the voltage waveform is the same as that of the current but leads it by 90°, since cos ω t = sin (ω t + 90°).

Thus the voltage in an inductive circuit leads the current by 90°.

To remember which waveform leads which in either a capacitive or an inductive circuit, the following mnemonic is useful.

CIVIL
.e in a capacitive circuit (C) the current (I) leads the voltage (V)

CIV..
in an inductive circuit (L) the voltage (V) leads the current (I)

..VIL
The phasor representation of voltage and current is given in fig 16.4.
Here the current is taken as the reference and the voltage leads it by 90°

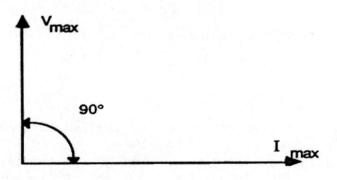

Fig 16.4

LAB 4

SEMICONDUCTOR DEVICES

– SECTION A SEMICONDUCTOR DIODES

THE SEMICONDUCTOR DIODE

OBJECTIVES

1. Ability to recognise diodes in various physical forms.

2. Ability to determine the diode polarity and to understand the need for correct connection.

3. To obtain knowledge of the forward voltage/current characteristic and the conduction voltage for germanium and silicon types.

EQUIPMENT REQUIRED

Qty   Apparatus

1   Electricity & Electronics Constructor, EEC470

1   Basic Electronics, Kit 2, EEC472

1   Power supply unit

0 to 20V variable d.c., regulated

(e.g Feedback Power Supply PS445)

2   Multimeters, or.
1   5V d.c. voltmeter and

1   High Impedance (20 kΩ/V) 0-3V d.c. voltmeter

PREREQUISITE ASSIGNMENTS

Assigment 1

KNOWLEDGE LEVEL

Before working this assignment you should:

●   Know the operation of series d.c. circuits

A Semiconductor Junction Diode (or just Diode) is made from a piece of P-type and a piece of N-type semiconductor joined together. See fig 2.1.

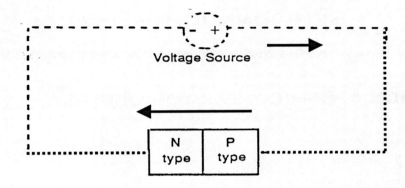

Fig 2.1 Function Diode

If a voltage (potential difference) is applied across the two terminals, the Diode will conduct electricity. The amount of current that flows depends upon the size and polarity of the applied voltage.

The Diode is represented in circuits by the symbol shown in fig 2.2.

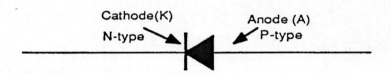

Fig 2.2 Diode Symbol

Find and examine the two Diodes supplied in the kit. They should appear as in fig 2.3.

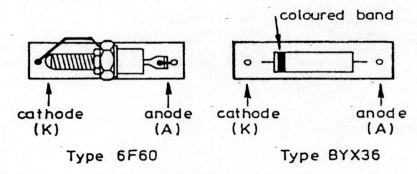

Fig 2.3 Two Types of Diode

The Diode with the metal body (type 6F60) can handle larger currents and power than the other one.

1   Which side of a diode should be connected to the positive voltage supply to make it conduct current?

2   When the diode was connected the opposite way round was the current?

   a) slightly smaller

   b) much smaller

   or   c) too small to measure

When a diode is connected so as to conduct it is **FORWARD BIASED**.

When a diode is connected so as NOT to conduct, it is **REVERSE BIASED**

Fig 2.8 shows the two methods of connecting diodes.

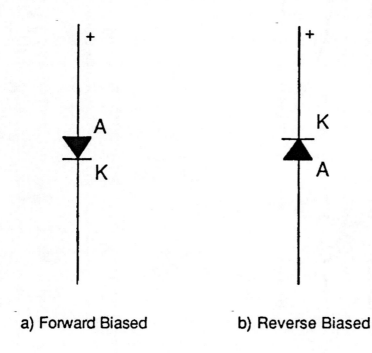

a) Forward Biased   b) Reverse Biased

Fig 2.8 Diode Bias

A knowledge of the conduction characteristic when a diode is forward biased is very important and is the subject of the rest of this Assignment.

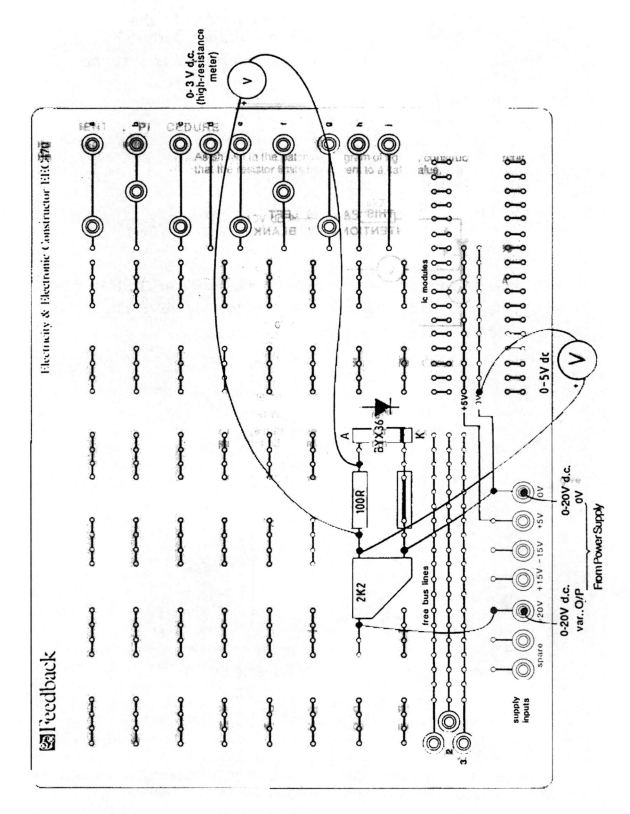

Fig 2.9

The Characteristics of Forward Biased Diodes

As shown in the patching diagram of fig 2.9, construct the circuit of fig 2.10. The 2.2kΩ potentiometer will provide fine control over the applied voltage.

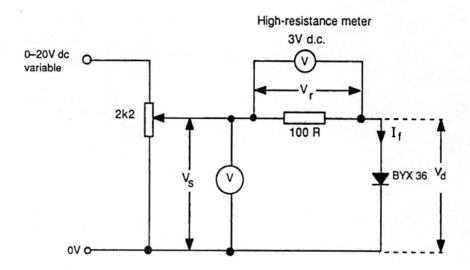

Fig 2.10 Test Circuit

NOTE: $V_d = V_s - V_r$ and $V_r = I_f \times 100$

$$\therefore I_f = \frac{V_r}{100} \, A$$

$$= \frac{V_r}{100} \times 1000 \, mA$$

$$\therefore I_f = 10V_r \, mA$$

Copy the results table as shown in fig 2.11, reproduced at the end of this assignment, for your results.

Turn the potentiometer to zero; fully clockwise.

Switch on the power supply and adjust it to supply 20V.

Adjust the potentiometer to give a voltage of 1V on the voltmeter showing V_s.

Now use the power supply variable control to set V_s to:

0, 0.1V, 0.2V, etc, up to 1.0V.

Note V_r for each setting and enter it in your table.

Now, with the power supply variable control set to supply 20V, use the potentiometer to set V_s to:

1.5V, 2.0V, 2.5V and 3.0V.

Again enter the values of V_r in your table.

Calculate V_d and I_f as shown in fig 2.11 and enter these also in the table.

Prepare a graph like Fig 2.12 on which to plot your results
Plot V_d against I_f on the axes of the graph.

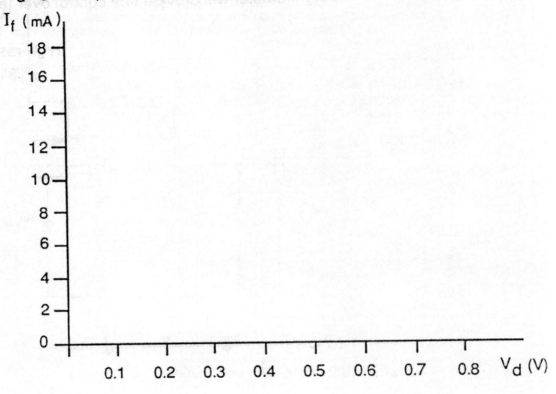

Fig 2.12 Diode Forward Characteristics

3 At what approximate value of V_d does the current I_f begin to rise noticeably?

4 Does V_d rise much above this value for larger values of I_f?

98

Both BYX36 and 6F60 diodes are made of Silicon and the forward conduction voltage of about 0.6V is typical of silicon junctions.

Also typical of silicon diodes is the very small reverse current.

Some diodes are made of Germanium and these have a smaller conduction voltage of about 0.2V but they also pass greater reverse currents.

The 6F60 diode passes a greater reverse current than BYX36. This is because 6F60 is designed for much larger forward currents — up to 6A average. At the low voltage used in this experiment the reverse amount will still be very small.

Diodes can withstand high reverse voltage but will eventually break down at some voltage and may be irreparably damaged. Type 6F60 can take the higher voltage of 600V compared with 150V for BYX36.

Diodes have very many applications at many different power, voltage and current levels. A very important application is the production of direct voltage from alternating voltage and this is dealt with in Assignments 3 and 4 which cover Rectification.

SUMMARY

In this assignment you have learnt that:

1. A diode conducts when its anode is positive relative to its cathode, and does not conduct when the voltage is reversed.

2. Diodes have different shapes and sizes according to their voltage, current and power ratings.

3. Silicon diodes have a conduction voltage of about 0.6V whereas Germanium diodes have one of about 0.2V.

4. The forward characteristic of a diode is not a straight line through zero but looks like fig 2.13.

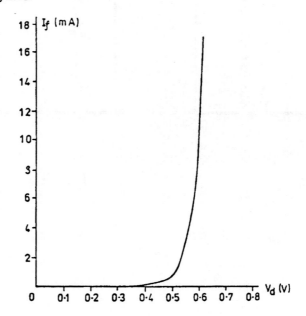

Fig 2.13 Typical Silicon Diode – Forward Characteristic

V_s (V)	V_r (V)	$V_d = V_s - V_r$ (V)	$I_f = 10V_r$ (mA)
0			
0.1			
0.2			
0.3			
0.4			
0.5			
0.6			
0.7			
0.8			
0.9			
1.0			
1.5			
2.0			
2.5			
3.0			

Fig 2.11

– SECTION B HALF-WAVE RECTIFICATION

HALF-WAVE RECTIFICATION

OBJECTIVES

1 To learn to recognise a half-wave rectified sinusoidal voltage.

2 To understand the term 'mean value' as applied to a rectified waveform.

3. To understand the effect of a reservoir capacitor upon the rectified waveform and its mean value.

EQUIPMENT REQUIRED

Qty	Apparatus
1	Electricity & Electronics Constructor EEC470.
1	Basic Electronics Kit EEC472.
1	Power supply unit. A.C. supply; 20Vrms; 50 or 60Hz (isolated from other supplies). (e.g Feedback Power Supply PS445)
1	Multimeter or 50V d.c. voltmeter.
1	Oscilloscope.

PREREQUISITE ASSIGNMENTS

Assignment 2

KNOWLEDGE LEVEL

Before working this assignment you should:

- Know the operation of transformers.
- Know the operation of series and parallel ac circuits
- Know how to use an oscilloscope.

101

INTRODUCTION

In Assignment 2 you found that a diode conducts current in one direction (from anode to cathode) but not in the reverse direction.

A widely used application of this feature is the conversion of alternating voltages to direct voltages (fig 3.1). This assignment studies the simplest circuits for achieving this conversion, which is called *RECTIFICATION* or, in some cases, *DETECTION*.

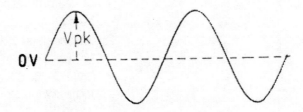

A Sinusoidal Alternating Voltage

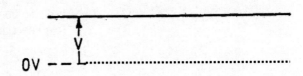

A Direct Voltage

Fig. 3.1

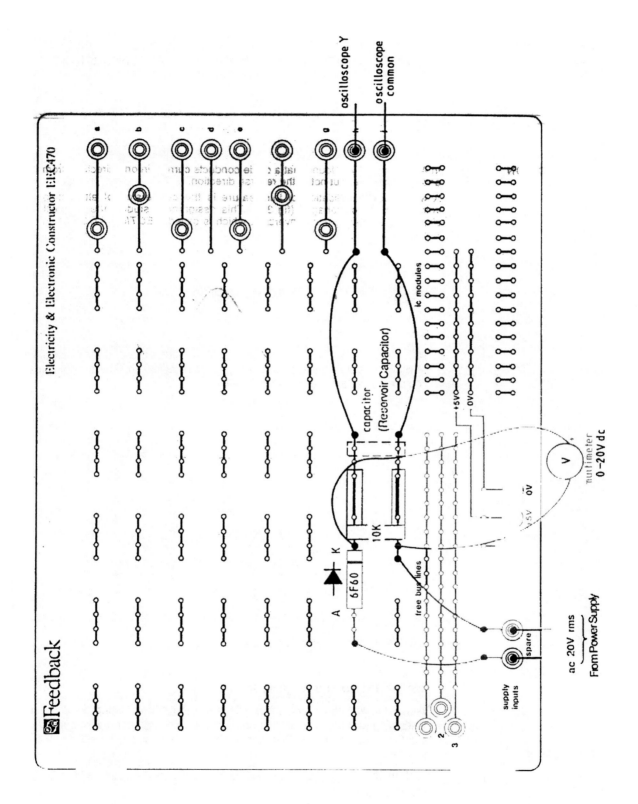

Fig 3.2

EXPERIMENTAL PROCEDURE

**Simple Half-wave
Rectification**

As shown in the patching diagram of fig 3.2, construct the circuit of fig 3.3.

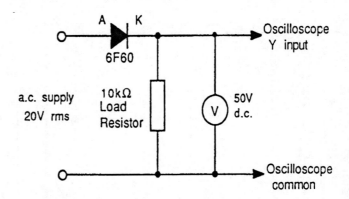

Fig 3.3 Half-wave Rectification

Switch on the oscilloscope and the sinusoidal supply.

With the oscilloscope d.c. coupled adjust the time-base and the Y amplifier sensitivity to obtain a steady trace of about 4cm vertical and 5ms/cm horizontal. You should see a waveform as in fig 3.4.

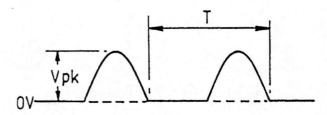

Fig 3.4 Half-wave Rectified Waveform

Measure and record time T and peak voltage V_{pk}.

Sketch the waveform and label it to show the periods when the diode is conducting and those when it is not. Time T depends upon the frequency of your power supply. For a 50Hz supply it should be 20ms and for 60Hz it should be 17ms.

Confirm this. V_{pk} should be very nearly equal to the peak voltage of the alternating supply.

1 *Why will Vpk not be exactly equal to this voltage?*

2 *How much will it differ?*

The waveform of fig 3.4 goes only positive relative to zero volts. If you connect a d.c. voltmeter across the output as in fig 3.2, the mechanical inertia of the meter will not allow the needle to respond to the rapid voltage changes. Instead, it indicates the *MEAN* voltage of the waveform.

The mean value of a half-sinusoid can be shown by geometry to be:

$$\frac{V_{pk}}{\sqrt{2}}$$

But at every half-cycle the voltage is zero. The mean value of the waveform, therefore, is:

$$\frac{1}{T}\left[\frac{V_{pk}}{\sqrt{2}} \times \frac{T}{2} + 0 \times \frac{T}{2}\right] = \frac{V_{pk}}{2\sqrt{2}}$$

Note the mean voltage indicated by the voltmeter, and compare it with $0.35 V_{pk}$.

3 *The mean voltage you obtain is positive relative to zero. How could you obtain a negative voltage?*

Confirm your answer by experiment.

The Effect of a Reservoir Capacitor

Very often when rectifying an alternating voltage, we wish to produce a steady direct voltage free from variations of the sort observed in fig 3.4. One way of doing this is to connect a capacitor in parallel with the load resistor as in fig 3.5.

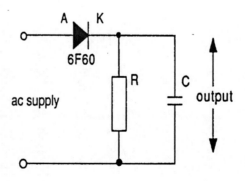

Fig 3.5 Half-wave Rectifier with Reservoir Capacitor

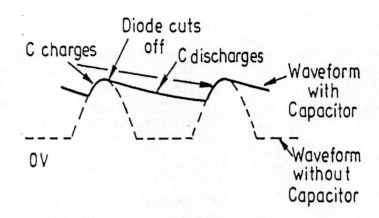

Fig 3.6 The Effect of a Reservoir Capacitor

The capacitor C (usually called the reservoir capacitor) becomes charged-up by the current through the diode during the positive half-cycle. Then, when the supply voltage starts to reduce again, the capacitor keeps the output voltage high and the diode cuts off. Capacitor C then discharges through R until the next positive half-cycle occurs.

Now add a capacitor of 1μF (shown dotted in the patching diagram of fig 3.) to your circuit.

Observe the output waveform on the oscilloscope and note the value of the peak-to-peak variations in voltage. Note also the new mean voltage on the voltmeter.

Question

4 *Is the new mean voltage greater or less than it was before?*

Now replace the 1μF capacitor by a much larger value of 22μF, making sure to connect the + side of the capacitor to the diode cathode (the capacitor is electrolytic and MUST be connected in the correct polarity) and answer the following questions.

Questions

5 *The variations on the rectified waveform are called RIPPLE. Is the ripple now less than or more than it was with the lower value capacitor?*

6 *Is the mean rectified voltage now greater or less?*

When rectification is used to provide a direct voltage power supply from an alternating source, the ripple is an undesirable feature. For a given capacitor value, a greater load current (smaller load resistor) discharges the capacitor more and so increases the ripple obtained. Fig 3.7 shows this.

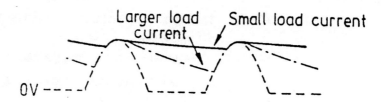

Fig 3.7 The Effect of Load Current

Several methods are available to reduce ripple:

1. Larger capacitors, the uses of which are limited due to cost and size, and also because large capacitors can require very large charging currents to be supplied through the diode.

2. Electronic stabilisation This reduces ripple as well as keeping the output voltage steady when the load or input voltage changes.

3. Full-wave rectification. With this, the ripple is much reduced as every half-cycle of the input, instead of every other half-cycle, contributes to the rectified output.

In fig 3.8 it can be seen that capacitor charging occurs at half the previous interval and the amount of discharge for a given load current is therefore less.

Fig 3.8 Full-wave Rectification

Assignment 4 deals with methods of achieving full-wave rectification.

When diodes are used for detection purposes in the reception of modulated radio signals, quite different considerations apply. These cannot be discussed in detail here but Feedback's manual ACS2956, Analogue Communications Systems, will provide full information on this application.

SUMMARY

In this assignment you learnt that:

1. A simple diode circuit can convert an alternating voltage to a direct voltage.

2. The mean value of the rectified voltage can be increased by using a reservoir capacitor across the load.

3. A half-wave rectified voltage gives appreciable ripple which however, can be reduced by several means.

EXERCISE

A half-wave rectifier, as in fig 3.9, produces a certain amplitude (from peak-to-peak) of ripple.

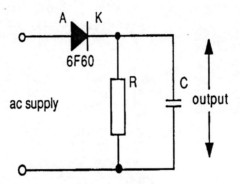

Fig 3.9 Exercise Circuit

If the load resistor is reduced to half of its original value, what increase in capacitor value will restore the ripple to the same value as before?

Confirm your answer by practical experiment, starting with:

$$R = 10k\Omega \text{ and } C = 22\mu F.$$

– SECTION C FULL-WAVE RECTIFICATION

FULL-WAVE RECTIFICATION

OBJECTIVES

1 Ability to recognise a full-wave rectified waveform, with and without a reservoir capacitor.

2 Understand the working of a diode bridge circuit as a full-wave rectifier and its advantage over half-wave rectification.

3 Awareness of the two-diode method of obtaining full-wave rectification.

EQUIPMENT REQUIRED

Qty Apparatus

1 Electricity & Electronics Constructor EEC470.

1 Basic Electronics Kit EEC472.

1 Power supply unit. A.C. supply; 20Vrms; 50 or 60Hz. (isolated from other supplies). (e.g Feedback Power Supply PS445)

1 Multimeter or

1 Voltmeter 50V d.c. .

1 Oscilloscope.

PREREQUISITE ASSIGNMENT

Assignment 2.

KNOWLEDGE LEVEL

Before working this assignment you should:

● Know the operation of series and parallel a.c. circuits.

● Know how to use an oscilloscope.

INTRODUCTION

At the end of Assignment 3 we discussed ways of reducing the ripple or voltage variation on a rectified direct voltage. One of these was to use every half-cycle of the input voltage instead of every other half-cycle.

A circuit which allows us to do this is shown in fig 4.1, and is known as the **DIODE BRIDGE**.

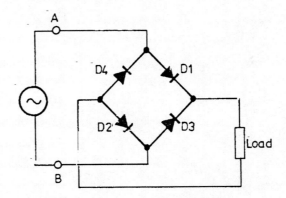

Fig 4.1 A Diode Bridge Rectifier

During the positive half-cycle of the supply 'A' is more positive than 'B'. Diodes D1 and D2 therefore conduct while diodes D3 and D4 are reverse-biased. The current flows as shown in fig 4.2.

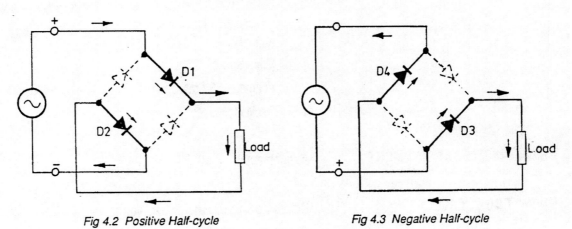

Fig 4.2 Positive Half-cycle   *Fig 4.3 Negative Half-cycle*

During the negative half-cycle the current flow is as represented by fig 4.3
In each case the current in the load is in the same direction.

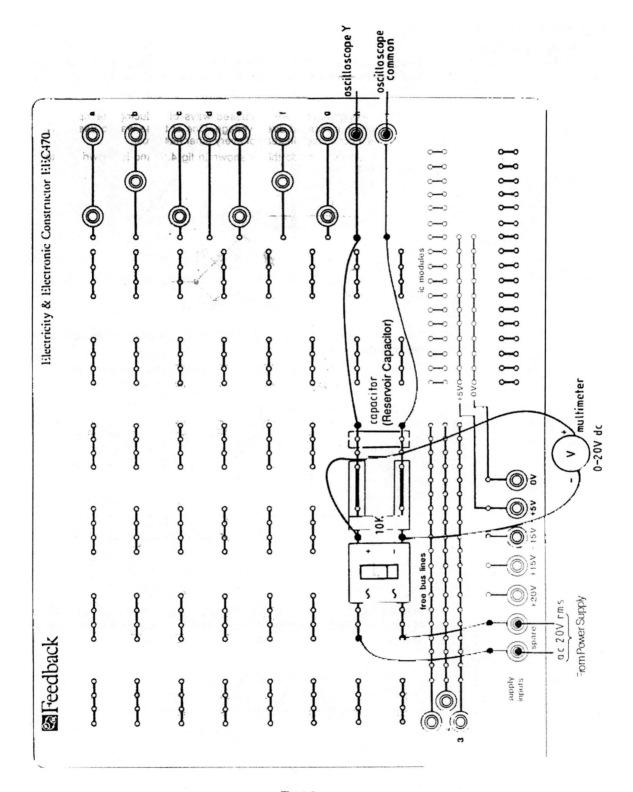

Fig 4.6

EXPERIMENTAL PROCEDURE

**A Bridge Rectifier
with Resistive Load**

Select the Bridge Rectifier from the component kit. It appears as in fig 4.4a and fig 4.4b in the circuit, showing how the rectifier terminals are labelled.

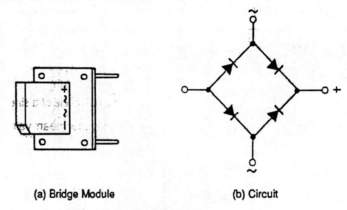

(a) Bridge Module   (b) Circuit

Fig 4.4 Bridge Rectifier

The terminals labelled + and - are so called because these are the polarities that will exist across the load.

Construct the circuit of fig 4.5 as in the patching diagram of fig 4.6.

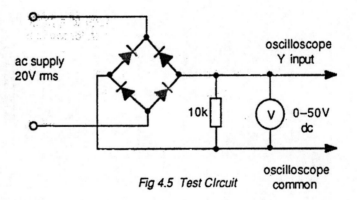

Fig 4.5 Test Circuit

With the oscilloscope d.c. coupled, adjust the controls to obtain a steady trace of about 4cm vertical and 5ms/cm horizontal. You should observe a waveform as in fig 4.7. Time 'T' will be 10ms for 50Hz supply, and 8.5ms for 60Hz.

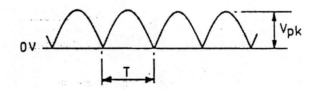

Fig 4.7 Full-wave Rectified Waveform

Note the value of Vpk and also the mean value of output voltage indicated on the voltmeter. Compare these figures with those obtained in Assignment 3.

Questions

1 *Should V_{pk} be the same as it was for a half-wave rectifier? Does your observation confirm your answer?*

2 *How does the mean value compare with that found for half-wave rectification?*

As the mean value of a half-cycle of a sine wave is $\dfrac{V_{pk}}{\sqrt{2}}$, and every half-cycle is present, this should be the mean value measured. Confirm this from your readings.

The Effect of a Reservoir Capacitor

Add a 1μF capacitor in parallel with the load resistor and note the new mean value and the peak-to-peak ripple amplitude of the rectified waveform. Compare these figures with those obtained in Assignment 3 for the same load and capacitor values .

PRACTICAL CONSIDERATIONS AND APPLICATIONS

The alternating input voltage to a rectifier is usually obtained from the main supply through a transformer, for two reasons:

1. To obtain the desired voltage by choice of the transformer ratio.

2. To provide isolation from the main supply for safety reasons.

Fig 4.8 shows such an arrangement with a bridge rectifier.

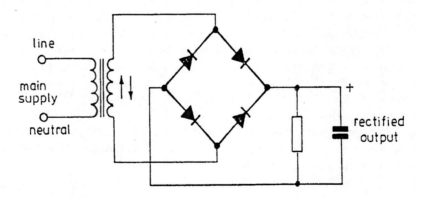

Fig 4.8 Transformer-fed Bridge Rectifier

In this figure, although the load current is always in one direction, the current in the transformer secondary is alternating.

Fig 4.9 shows another method of full-wave rectification, using a centre-tapped transformer winding and two diodes.

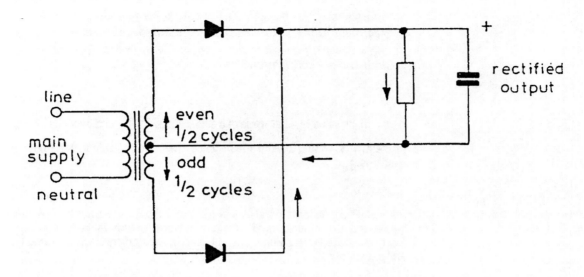

Fig 4.9 Full-wave Rectifier using Two Diodes

The arrows show how current flows on alternate half-cycles. The value of the output waveform is exactly the same as that for a bridge circuit provided each half of the transformer windings has the same rms voltage as the whole of the winding in fig 4.8.

The circuit saves two diodes, but increases the cost of the transformer. In fig 4.9 each half-secondary winding must have the same voltage rating as the single secondary of fig 4.8. Suppose the half-secondaries were wound with wire of half the cross-sectional area, so as to fit the two into the same space as the one secondary of fig 4.8, and use the same amount of copper. Each half-secondary would then have twice the resistance.

The current flows in each half-secondary only on alternative half–cycles, but would generate twice the I^2R loss in the active cycle.

Each half-secondary would thus develop as much heat as the single secondary of fig 4.8, i.e twice as much for both. A larger transformer would therefore be required to avoid excessive heating. Its greater cost would usually outweigh the cost of the two diodes saved.

In full-wave rectification the basic repetition rate of the ripple is twice that of the supply (e.g 100Hz for a 50Hz supply). In half-wave the frequency is the same as the supply frequency. This is often useful as an indication that one half of a bridge or full-wave rectifier is faulty.

114

SUMMARY

In this assignment you have learnt that:

1. A bridge full-wave rectifier gives a greater mean value and less ripple for a given load and reservoir capacitor than a half-wave rectifier.

2   The alternative full-wave circuit using a centre-tapped transformer and two diodes is less efficient than the bridge circuit because it requires a bigger transformer for a given output power.

EXERCISE

Fig 4.10 shows the discharge curve for a reservoir capacitor in half-wave and full-wave rectification, for the same load and capacitor values.

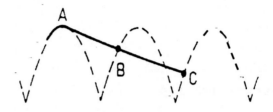

A— Start of discharge

B— End of discharge; full wave

C — End of discharge; half wave

Fig 4.10

A capacitor discharges into a resistor in an exponential fashion, that is with a rate of discharge that reduces as the discharge progresses.

With this in mind, would you expect the peak-to-peak ripple in full-wave to be:

(a) $\frac{1}{2}$ that in half-wave

(b)   less than $\frac{1}{2}$

(c)   more than $\frac{1}{2}$?

Explain your answer and confirm it by reference to measurements made in Assignments 3 and 4 for similar load conditions.

– **SECTION D** Zener Diode Circuits

THE ZENER DIODE

OBJECTIVES

1 Ability to recognise zener diodes in various physical forms and to distinguish them from rectifying diodes.

2 Understand the constant-voltage characteristic of a reverse-biased zener diode.

3 Understand the use of a zener diode in a simple voltage regulator circuit.

EQUIPMENT REQUIRED

Qty	Apparatus
1	Electricity & Electronics Constructor EEC470.
1	Basic Electronics Kit EEC472.
1	Power supply unit 0 to 20V variable d.c,. regulated
	(e.g Feedback Power Supply PS445)
3	Multimeters, or
1	Voltmeter, 20V d.c and
1	Ammeter, 100mA d.c and
1	Ammeter, 1A d.c.

PREREQUISITE ASSIGNMENTS Assignment 2.

KNOWLEDGE LEVEL

Before working this assignment you should:

● Know what is meant by internal resistance and the effect it has on terminal voltage.

In Assignment 2 you found that a reverse-biased diode passes negligible current. You also learnt that it will eventually suffer breakdown and damage if the reverse voltage is made too high. See fig 5.1.

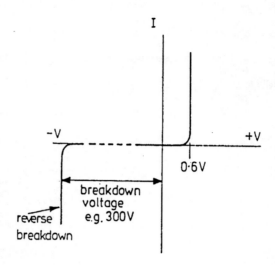

Fig 5.1 Reverse Breakdown of a Diode

Zener diodes are specially constructed to break down at controllable voltages and to do so without damage to the device. As we shall see, this feature can be put to good use.

Two Zener diodes are contained in the EEC472 Kit. They are types BZY95C10 and BZY88C7V5 and are shown in fig 5.2 with the standard circuit symbol.

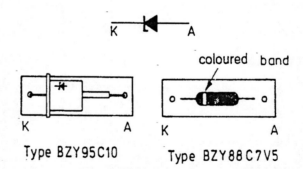

Fig 5.2 Zener Diodes and Symbol

Zener diodes look very similar to rectifier diodes and the terminal names and identification methods are the same. The larger types, such as BZY95, have greater power and current capacities.

The two types of diode can usually be distinguished only by their type numbers. For Zener diodes these often, but not always, contain the letter Z.

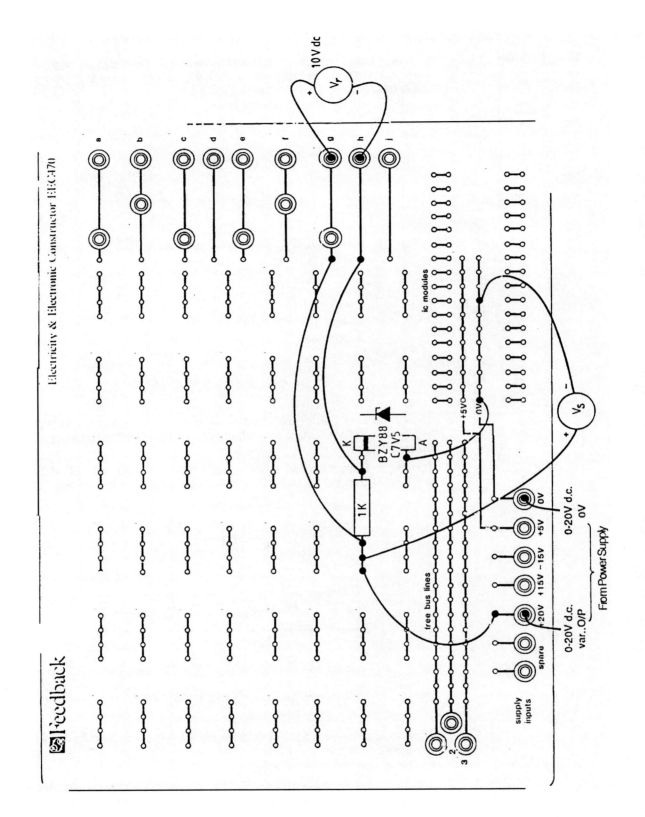

EXPERIMENTAL PROCEDURE
The Zener Diode
Reverse Characteristics

As shown in the patching diagram of fig 5.3, construct the circuit of fig 5.4

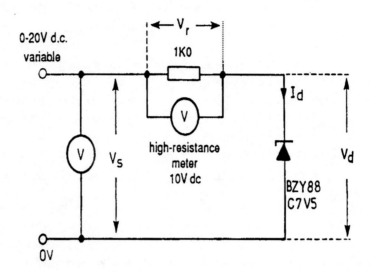

Fig 5.4 Test Circuit

The method of obtaining the voltage-current characteristic is the similar to that of Assignment 2 but notice that the Zener diode is reverse-biased. Using the power supply variable control, set V_s to the values given in fig 5.5.

For each value record V_r, then calculate:

$$V_d = V_s - V_r \quad \text{and} \quad I_d = \frac{V_r}{1000} = V_r\,mA$$

Copy the results table as shown in fig 5.5, reproduced at the end of this assignment, and enter your results.

Prepare a graph like fig 5.6 and plot V_d against I_d

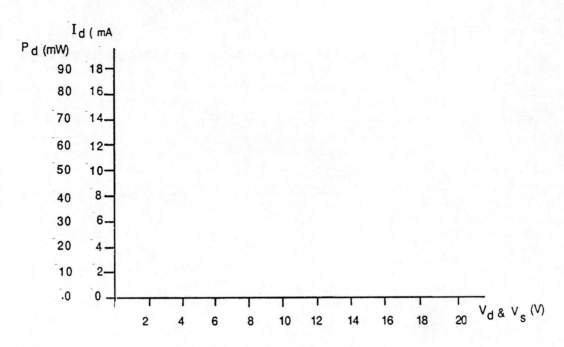

Fig 5.6 The Zener Diode Characteristic

Questions

1 . Describe the characteristic in your own words.

2 . The nominal voltage of the BZY887V5 is 7.5V. Does your graph agree with this exactly? If not, can you suggest a reason for any difference?

3 . Why is the series resistor in fig 5.4 necessary?

Calculate the power dissipated in the diode for each value of V_d and I_d, and enter it into the last column of your table.

$$P_d = V_d \times I_d \text{ mW}$$

Plot P_d against V_d on your graph.

Question

4 . The maximum allowable power dissipation of type BZY88 is 400mW. Does your maximum value of P_d approach this limit?

120

V_s (V)	V_r (V)	$V_d = V_s - V_r$ (V)	$I_d = V_r$ (mA)	$P_d = V_d \times I_d$ (mW)
0				
2				
4				
6				
6.5				
7.0				
7.5				
8.0				
8.5				
9.0				
10.0				
15.0				
20.0				

Fig 5.5

LAB 5

SEMICONDUCTOR DEVICES

– SECTION A SEMICONDUCTOR TRANSISTORS

TRANSISTOR FAMILIARISATION

OBJECTIVES

1. Ability to recognise transistors in various physical forms and to identify their terminals.

2. Understanding of the basic construction of PNP and NPN transistors.

3. Understanding of junction biasing and the direction and magnitude of current flows.

EQUIPMENT REQUIRED

Qty	Apparatus
1	Electricity & Electronics Constructor EEC470.
1	Basic Electronics Kit EEC472.
1	Power supply unit +5V and –15V variable d.c., regulated (e.g Feedback Power Supply PS445)
2	Multimeters, or
1	Microammeter,100µA d.c. and
1	Voltmeter, 3V d.c.

PREREQUISITE ASSIGNMENTS

Assignment 2

KNOWLEDGE LEVEL

Before working this assignment you should:

● Know the operation of parallel d.c. circuits

INTRODUCTION

Transistors are three-terminal devices constructed in the form of two semiconductor junctions, rather like two junction diodes. Fig 6.1 shows the two types, NPN and PNP, governed by the physical arrangement of the P- and N-type semiconductor materials.

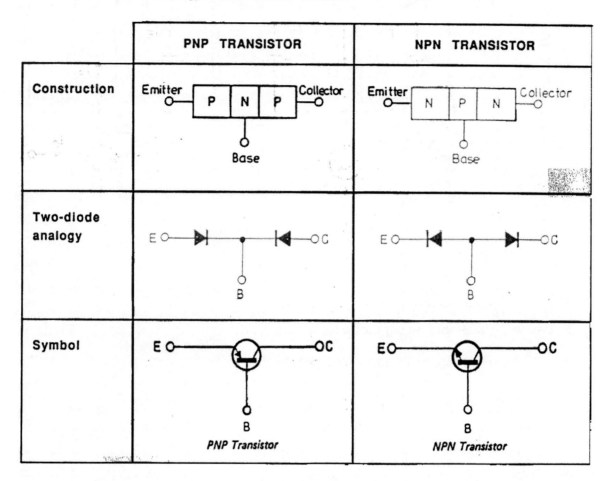

	PNP TRANSISTOR	NPN TRANSISTOR
Construction	Emitter — P N P — Collector, Base	Emitter — N P N — Collector, Base
Two-diode analogy	E ▷ ● ◁ C, B	E ◁ ● ▷ C, B
Symbol	E — C, B, *PNP Transistor*	E — C, B, *NPN Transistor*

Fig 6.1 Two Types of Transistor

Each of the PN junctions in this diagram behaves individually like the simple diode you studied in Assignment 2, but when joined together in this way, the behaviour is very different.

In normal use the **EMITTER-BASE** diode is forward-biased and behaves almost exactly like an independent diode. The **COLLECTOR-BASE** diode, however, is reverse-biased and normally you would expect if to pass no current. But if the E-B diode is conducting forward current, this influences the reverse-biased C-B diode and causes it to pass almost as much reverse current.

Fig 6.2 shows this for PNP and NPN types. The small difference current flows in the base circuit.

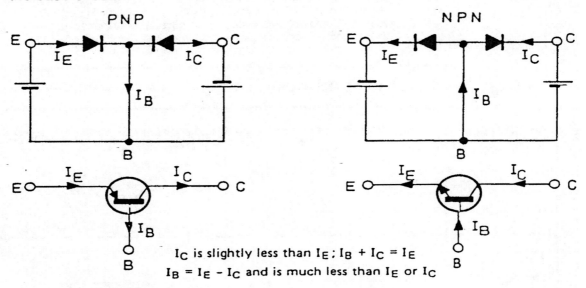

I_C is slightly less than I_E; $I_B + I_C = I_E$
$I_B = I_E - I_C$ and is much less than I_E or I_C

Fig 6.2 Transistor Current Flow

The ratio $\dfrac{I_C}{I_E}$ is usually called h_{fb}.

Because I_C is almost as big as I_E, h_{fb} is nearly 1

$$\therefore \quad \frac{I_C}{I_E} = h_{fb} = \text{nearly 1 (e.g. 0.99)}$$

The ratio $\dfrac{I_C}{I_B}$ is usually called h_{fe}

$$I_B + I_C = I_E = \frac{I_C}{h_{fb}}$$

Thus $\quad I_C\left(\dfrac{1}{h_{fb}} - 1\right) \qquad = I_B$

and $\quad h_{fe} = \dfrac{I_C}{I_B} \qquad = \dfrac{1}{\dfrac{1}{h_{fb}} - 1} = \dfrac{h_{fb}}{h_{fb} - 1}$

if $\quad h_{fb} \qquad = 0.99,$

$\qquad\qquad h_{fe} \qquad = \dfrac{0.99}{0.01}$

$\qquad\qquad\qquad\qquad = 99.$

It is this large ratio between I_C and I_B that makes the transistor a useful amplifying device when connected so that I_B is derived from an input and I_C provides an output.

In the Assignment we shall first identify some actual transistors and then confirm the directions and magnitudes of currents, finding h_{fb} and h_{fe} in the process.

EXPERIMENTAL PROCEDURE

Transistor Identification

Select from the EEC472 kit the transistors BC107 and BCY70. Fig 6.3 illustrates these types.

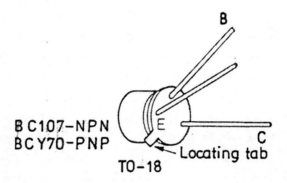

Fig 6.3 Typical Low-power Transistor (as supplied)

Transistors are made in many other physical forms. Fig 6.4 shows some other types you could have to recognise.

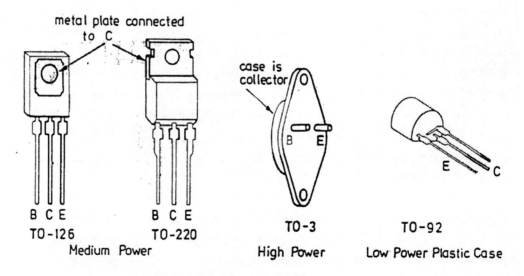

Fig 6.4 Different Transistor Styles

Make sure you can accurately identify the terminals of the transistors in the kit.

PRACTICAL CONSIDERATIONS
AND APPLICATIONS

The measurements you made on the BC107 transistor used a circuit in which the E and C terminals were biased with voltages relative to the base B. For this reason this circuit is called a **COMMON BASE** connection.

It is also possible to bias the junctions with voltages relative to the EMITTER or **COLLECTOR**, giving **COMMON EMITTER** and **COMMON COLLECTOR** connections as in fig 6.8.

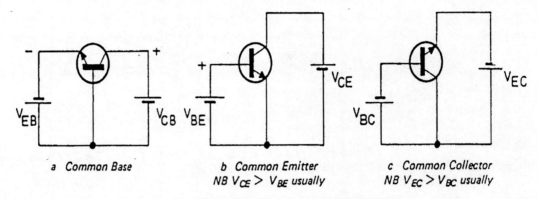

a Common Base

b Common Emitter
NB $V_{CE} > V_{BE}$ usually

c Common Collector
NB $V_{EC} > V_{BC}$ usually

Fig 6.8 Bias Arrangements for an NPN Transistor

In common-emitter, V_{CE} must be larger than V_{BE} to ensure that the C – B junction remains reverse-biased.

In common-collector, V_{EC} must be larger than V_{BC} to ensure that the E – B junction remains forward-biased.

These three connections have important differences in their responses to inputs. The common-emitter and common-collector circuits are the most important connections since the common-base is used only in special circumstances.

As with diodes, transistors can be made from Germanium instead of Silicon, but these are rarely used.

SUMMARY

In this assignment you have learnt that:

1. Transistors have two basic forms; the PNP and the NPN.

2. A transistor is similar to two diode junctions, one forward and one reverse-biased.

3. The base current is much smaller than either the emitter or collector current, which are themselves nearly equal.

4. There are three basic bias connections for a transistor.

– SECTION B Common-Emitter Circuit

THE COMMON–EMITTER TRANSISTOR CIRCUIT

OBJECTIVES

1   To familiarise the student with the common-emitter output (collector) characteristics.

2   To provide an understanding of the meaning and importance of OPERATING POINT and LOAD LINE.

EQUIPMENT REQUIRED

Qty Apparatus

1   Electricity & Electronics Constructor EEC470

1   Basic Electronics Kit EEC472

1   Power supply unit 0 to 20V variable d.c.regulated and +5V d.c. regulated.

(e.g Feedback Power Supply PS445)

3   Multimeters or

1   Microammeters 100µA d.c

1   Milliammeter 10mA dc

1   High resistance voltmeter 10V d.c.

PREREQUISITE ASSIGNMENTS

Assignment 6

KNOWLEDGE LEVEL

Before working this assignment you should:

●   Know the operation of a potential divider circuit.

Introduction

In Assignment 6 you learnt that the emitter and collector currents of a transistor are nearly equal. Also that the base current is the difference between them and is therefore much smaller.

You learnt too that the transistor junctions can be biased in three different ways, called 'common-base', 'common-emitter' and 'common-collector'. This assignment looks more closely at the common-emitter connection, using an NPN type of transistor.

Fig 7.1 reminds us of some of the results from Assignment 6.

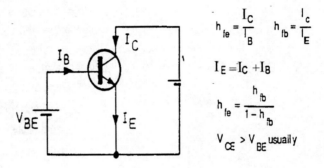

$$h_{fe} = \frac{I_C}{I_B} \qquad h_{fb} = \frac{I_c}{I_E}$$

$$I_E = I_C + I_B$$

$$h_{fe} = \frac{h_{fb}}{1 - h_{fb}}$$

$$V_{CE} > V_{BE} \text{ usually}$$

Fig 7.1 NPN Transistor Common-emitter Connection

If I_B is derived from an input signal and I_C is used to generate an output, then the ratio $\dfrac{I_C}{I_B} = h_{fe}$ represents the gain of the transistor in terms of the currents.

That is:- $h_{fe} = \dfrac{\text{OUTPUT CURRENT}}{\text{INPUT CURRENT}} = \text{CURRENT GAIN}$

What we now need is a graph that will tell us exactly how I_C, I_B and V_{CE} are related to one another. For instance we know that for the C-B junction to be reverse-biased $V_{CE} > V_{BE}$, but what happens if it is not?

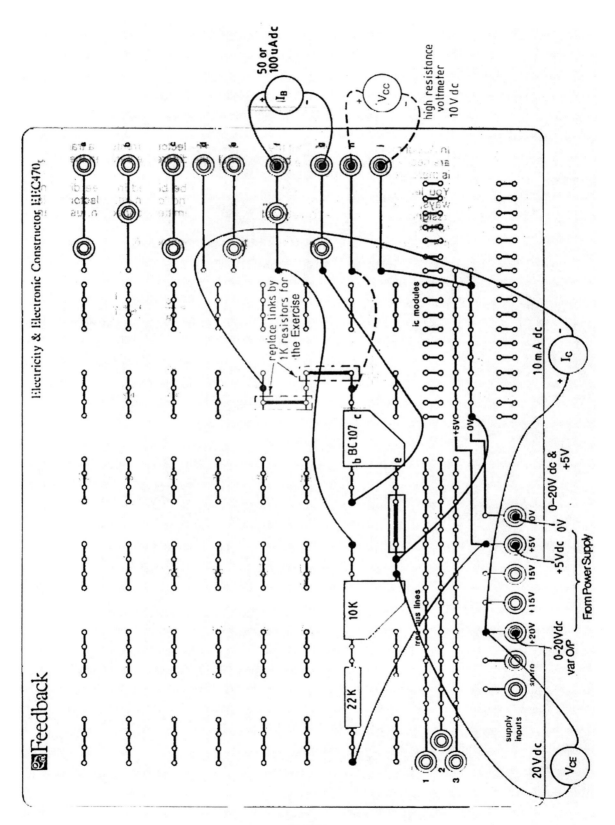

Fig 7.2

EXPERIMENTAL PROCEDURE

To find how I_C is
controlled by I_B and V_{CE}

Construct the circuit of fig 7.3.as shown in the patching diagram of fig 7.2,

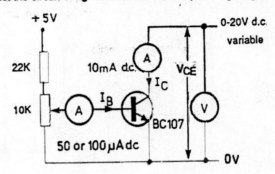

Fig 7.3 Test Circuit

Copy the results table as shown in fig 7.4, reproduced at the end of this
assignment, and prepare a graph as shown in fig 7.5, for your results.

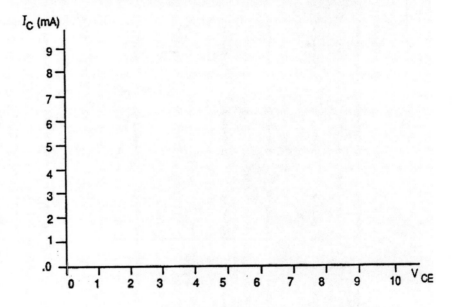

Fig 7.5 BC107 Collector Characteristics

Set V_{CE} to 0.5V, then use the potentiometer to adjust I_B to each value given
in the table of fig 7.4.

At each setting record I_C in the appropriate column. Then repeat for each other
V_{CE} value.

Plot I_C agaist V_{CE} for each value of I_B on your graph.

The graphs you now have are the 'output' or 'collector characteristics'.

130

1 *What happens to I_C when V_{CE} becomes less than 0.6V? What is the significance of this value?*

2 *What do you notice about the effect of V_{CE} upon I_C when V_{CE} is greater than about 1.0V?*

The answer to question 2 is very important. You should have concluded that I_C is very little affected by V_{CE}. Instead however, it is almost entirely controlled by I_B.

We say that the output circuit represents a **CONSTANT CURRENT SOURCE.**

If we wish to produce an output voltage, as we might in certain kinds of amplifier, this current may be passed into a resistor to generate a voltage.

Constructing a Load Line

Fig 7.6 shows a circuit in which the collector bias is applied through a load resistor.

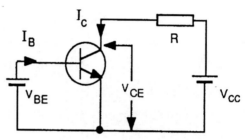

Fig 7.6 Addition of a Load Resistor

V_{CC} is the term now used for the bias voltage to distinguish it from V_{CE}, because when a current I_C is flowing these two will be different.

By Ohm's Law; $$V_{CE} = V_{CC} - I_C R.$$

Therefore, when $I_C = 0$, $$V_{CE} = V_{CC}$$

and when $V_{CE} = 0$ $$I_C = \frac{V_{CC}}{R}$$

Now, on your graph plot these two points V_{CE} and I_C for:

$I_C = 0$, $R = 2k\Omega$, and $V_{CC} = +10V.$

Then plot two or three more points for other values of I_C such as 2 and 5mA using the same values of R and V_{CC}.

Join up the points by a line.

Your line should be straight because the equation ($V_{CE} = V_{CC} - I_C R$) is a linear one; there being no square or cubic terms in it.

131

What you have now drawn is called a **LOAD LINE**. It shows how V_{CE} varies with I_C and in turn with I_B.

Construct a second load line for:

$$V_{CC} = 8V \text{ and}$$
$$R = 1k \text{ ohms.}$$

Copy the results table as shown in fig 7.7, reproduced at the end of this assignment, for your results, then complete the exercise.

EXERCISE

Fill in the gaps in your table by examination of your graph. Make estimates if an exact value cannot be obtained.

Now alter your constructed circuit to include a load resistor of 1kΩ and set V_{CC} = 8V. The circuit now looks like fig 7.8. Also see patching diagram fig 7.2.

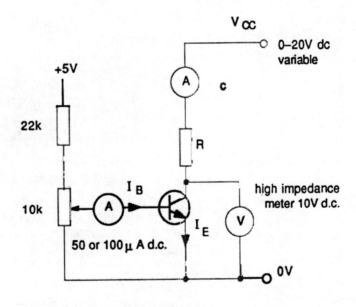

Fig 7.8 Test Circuit

Check your answers to items No. 1 and 4 in the table, by setting the given parameter and measuring the other two.

Now change V_{CC} and R to 10V and 2kΩ respectively and check your answers to items 2 and 3.

A circuit like fig 7.8 is widely used to amplify alternating signals. It is then necessary to set the initial value of V_{CE} to allow the output to vary both up and down. This is called **SETTING THE OPERATING POINT** and is what you have just done.

Very often the operating point will be set to V_{CE} which approximately equals $\dfrac{V_{CC}}{2}$ to allow the maximum possible swing in either direction.

Mark this point on each of the two loads lines on your graph (fig 7.5) and label the points as operating points.

Estimate the extreme values of V_{CE} for each load line if I_B varies by $\pm 10\mu A$ about the operating point value. Draw a thick line along the section of load line included within these limits. This represents a typical **OPERATING RANGE.**

Your graph should now look like fig 7.14.

PRACTICAL CONSIDERATIONS AND APPLICATIONS

Setting the operating point is an important matter in transistor amplifiers. In the Practicals above you have effectively done this by applying a voltage to the base-emitter junction through a resistor, as in fig 7.9.

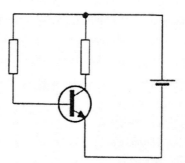

Fig 7.9 Simple Bias Circuit

Unfortunately this does not give a stable operating point in practice because temperature changes cause a considerable change in I_C for a given I_B. Also individual transistors of one type show some variation in characteristics.

Fig 7.10 shows a circuit which is much used to stabilise the operating point against such variations.

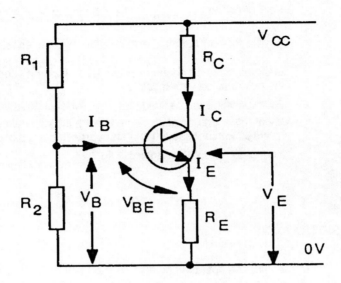

Fig 7.10 A Stabilised Bias Circuit

R_1 and R_2 set a voltage at the base, V_B, of say 4V and I_E is made to pass through resistor R_E. The emitter voltage, by Ohm's Law must be:

$$V_E = I_E \times R_E$$

and since V_{BE} is about 0.6V, then:

$$V_B = V_E + 0.6 = I_E R_E + 0.6$$

Thus, if $\qquad V_B = 4V, \quad V_E = 3.4V.$

Now if I_E increases, this causes V_{BE} to reduce and in turn reduces I_B and hence I_E. Thus the original change is counteracted. If I_E reduces the result is similar. The operating point has effectively been set by V_B and R_E, because:

$$I_C \cong I_E = \frac{(V_B - 0.6)}{R_E}$$

SUMMARY

In this assignment you have learnt that:

1. The collector or output characteristics for a common-emitter transistor can be used to predict I_C, given V_{CE} and I_B.

2. A load line can be constructed on the characteristic to show the effect of a resistor in the collector lead.

3. The load line can be used to determine a suitable operating point which can be set by adjustment of the base current.

4. The variation in base current determines the operating range.

EXERCISE

Use the value of h_{fe} found in Assignment 6 for the BC107 to estimate the peak-to-peak variation of V_{CE} in the circuit of fig 7.11 when I_B varies by ±2μA.

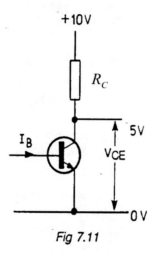

Fig 7.11

Also what mean level of I_B must you use to establish the operating point correctly?

(HINT: Variation of V_{CE} = R × variation of I_C)

$$V_{CE1} = V_{CC} - I_{C1}R_C$$
$$V_{CE2} = V_{CC} - I_{C2}R_C$$

$$\Delta V_{CE} = V_{CE1} - V_{CE2} = V_{CC} - I_{C1}R_C - (V_{CC} - I_{C2}R_C) = (I_{C2} - I_{C1})R_C = \Delta I_C R_C$$

Transistor Characteristics

The collector characteristics, with the two load lines, operating points and operating ranges, should appear as in fig 7.14 below.

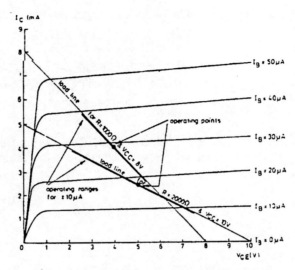

Fig 7.14 Typical BC107 Characteristics

Exercise

Assuming h_{fe} equals, say, 120;

The change in... $I_C = 120 \times 4\mu A = 0.48mA$

and in... $V_{CE} = (0.48 / 1000) \times 5000 = 2.4V$ pk-to-pk.

Then, change in V_{CE} = R x change in I_C

Thus: $R = \dfrac{\text{change in } V_{CE}}{\text{change in } I_C}$

$= \dfrac{2.4}{0.48 \times 10^{-3}}$

$= 5.33 \times 10^{-3} \Omega$

Also: $I_C = \dfrac{V_{CC} - V_{CE}}{R}$

$= \dfrac{10 - 5}{5 \times 10^3}$

$= 1 \times 10^{-3} A$

Then: $I_B = \dfrac{I_C}{h_{fe}}$

$= \dfrac{1.10^{-3}}{120}$

$\approx 8\mu A$

I_B	I_C (mA) for V_{ce} = ...(V)				
μA	0.5	1	2	5	10
0					
10					
20					
30					
40					
50					

Fig 7.4

No.	V_{CC} (V)	R (Ω)	V_{CE} (V)	I_B (μA)	I_C (mA)
1	8	1000		25	
2	10	2000			4
3	10	2000	5		
4	8	1000			5.5

Fig 7.7

137

LAB 6

TRANSISTOR AMPLIFIERS

- SECTION A: FIELD EFFECT TRANSISTORS
- SECTION B: BJT AND JFET AMPLIFIERS
- SECTION C: FREQUENCY RESPONSE AND NEGATIVE FEEDBACK NETWORK

SECTION A: ▬▬▬▬▬▬▬▬▬▬

FIELD EFFECT TRANSISTORS

THE FIELD EFFECT TRANSISTOR

OBJECTIVES

1 Understanding of the difference between bipolar and field-effect transistors.

2 Ability to distinguish between **JFET** and **MOSFET** types and between N and P channel construction.

3. Be able to recognise the basic characteristics of a JFET.

4. Know the principal advantages of FET's and some applications.

EQUIPMENT REQUIRED

Qty	Apparatus
1	Electricity & Electronics Constructor EEC470
1	Basic Electronics Kit EEC472
1	Power supply unit 0 to +20V variable d.c. and ±15V d.c. regulated (e.g Feedback Power Supply PS445)
1	Oscilloscope
3	Multimeters or
1	Voltmeter, 3V d.c. and
1	Milliammeter, 50µA/10 mA d.c. and
1	Voltmeter 50V d.c.
1	Function Generator 2V pk–pk at 1kHz (e.g. Feedback FG601)

KNOWLEDGE LEVEL

Before working this assignment you should
● Know how to use an oscilloscope.

Field-Effect Transistors (FET's) are made in various forms. One type, the Junction FET (JFET) has a construction quite similar to the UJT (Assignment 10) but works in a different way.

Fig 11.1 shows the construction, graphical symbol and physical appearance of a typical JFET.

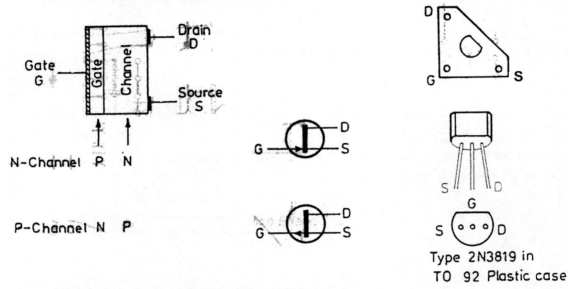

Fig 11.1 JFETs – Construction, Symbols, and Appearance.

As can be seen, the JFET has two forms; the N-channel and P-channel which are analogous to PNP and NPN in ordinary transistors. We shall look more closely at the N-channel type.

Fig 11.2 shows an N-channel JFET and its bias voltages.

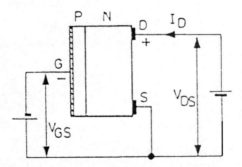

Fig 11.2 The Bias Arrangement for an N-channel JFET.

The **CHANNEL** is a resistive path through which voltage V_{DS} can drive a current I_D.

A voltage gradient is thus formed down the length of the channel, the voltage becoming less positive as we go from **DRAIN** to **SOURCE**. The PN junction thus has a high reverse bias at D and a lower reverse bias at S. This bias causes a 'DEPLETION LAYER', whose width increases with the bias.

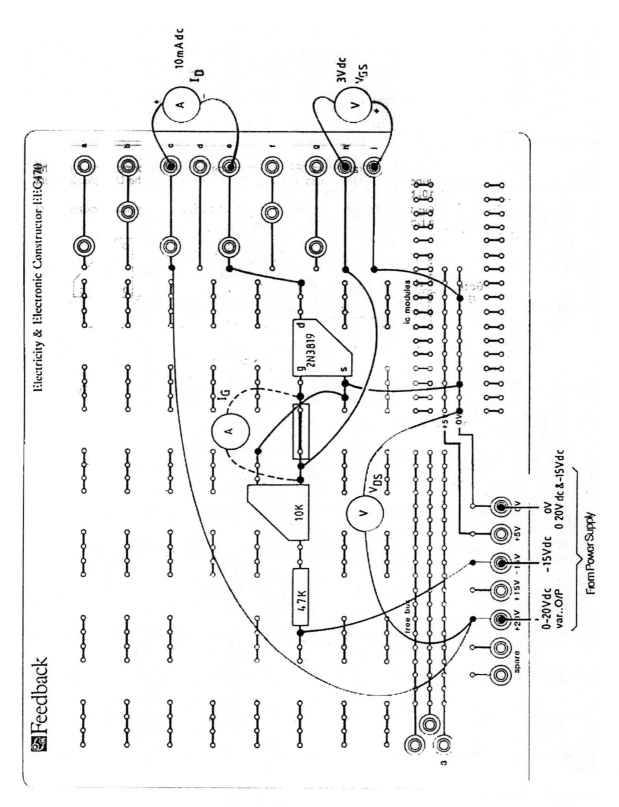

Fig 11.5

Depletion means a reduction of available electrons to carry current. If V_{GS} is made more negative, the depletion layer increases in width at all points. The values of V_{DS} and V_{GS} both influence the width of the depletion layer. This alters the effective channel resistance and hence I_D. Fig 11.3 shows this.

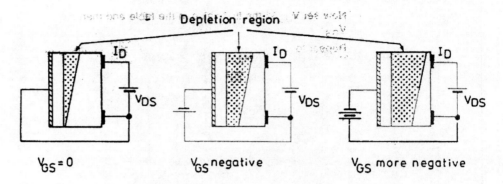

Fig 11.3 The Depletion Effect.

As V_{GS} increases negatively the channel is 'squeezed', reducing the current I_D. But the **GATE-CHANNEL** junction is like a reverse-biased junction diode and thus carries only a very small current. I_D is controlled by V_{GS} through a 'field effect'. Hence the name **FET**.

In the first Practical we shall see how V_{DS} and V_{GS} affect I_D.

EXPERIMENTAL PROCEDURE

**Characteristics of
an N-Channel JFET**   Construct the circuit of fig 11.4 to the patching diagram of fig 11.5.

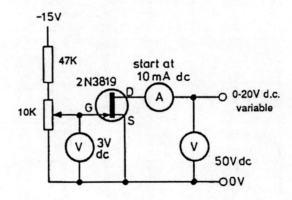

Fig 11.4 Test Circuit for a Typical N-channel JFET.

Set the potentiometer anti-clockwise and the variable d.c. voltage to zero. Switch on the power supply.

Copy the results table as shown in fig 11.6, reproduced at the end of this assignment, for your results.

Now set V_{DS} to the first value in the table and then read I_D for each value of V_{GS}.

Repeat for all the values of V_{DS} in the table, recording the corresponding I_D values.

Prepare a graph like fig 11.7.

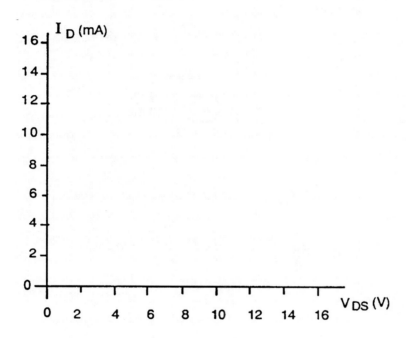

Fig 11.7 JFET Drain Characteristics.

Plot the results from your table onto your graph, drawing one curve of I_D against V_{DS} for each value of V_{GS}.

Study your graphs and answer the following questions:

1 **Above which values of V_{DS} is I_D almost unaffected by V_{DS} when $V_{GS} = 0$?**

2 **For a given value of V_{DS}, (say 10V), do equal changes of V_{GS} cause equal changes of I_D?**

Understand what your answer implies.

Now go back to your circuit and set V_{DS} to 10V and V_{GS} to −1.0V. Then alter the circuit to place the ammeter in place of the link in the gate lead as in fig 11.8. and try to measure I_G.

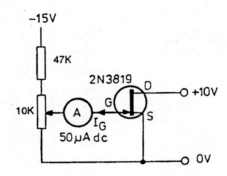

Fig 11.8 Measuring Gate Current.

3 *Can you now measure I_G or is it too small?*

144

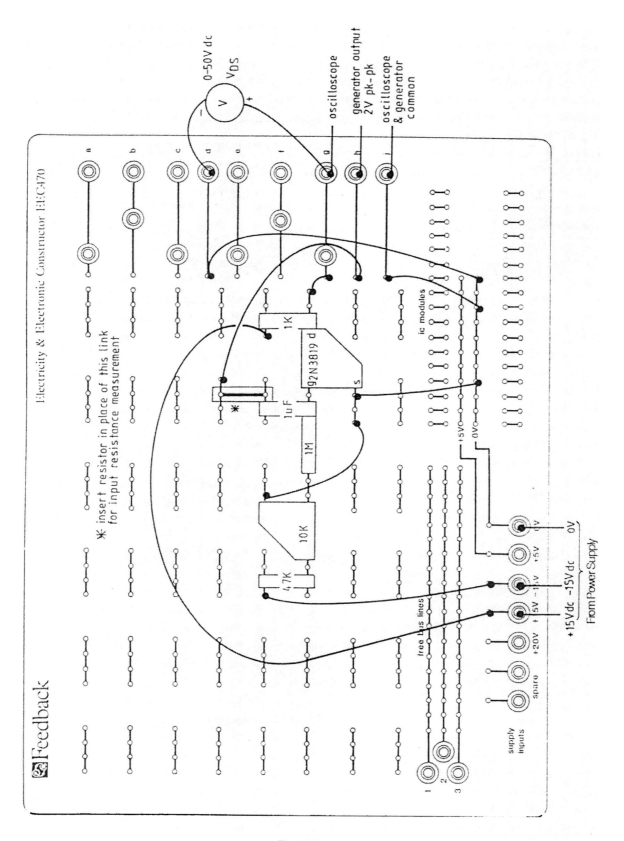

Fig 11.9

145

A JFET Amplifier

An FET can be used to amplify signals in a manner similar to a transistor in common-emitter connection. In this case we call it **COMMON-SOURCE**.

To obtain an output voltage we insert a load resistance in the drain lead, the effects of this being represented on the characteristic by a load line.

Fig 11.10 shows a practical amplifier circuit with a typical characteristic and load line.

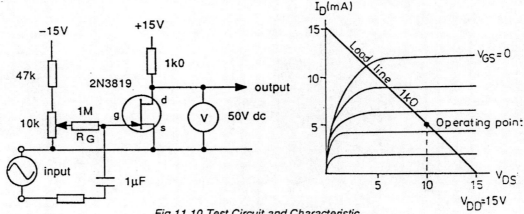

Fig 11.10 Test Circuit and Characteristic.

As shown in the patching diagram of fig 11.9, construct the circuit of fig 11.10.

On your graph draw a load-line from $V_{DD} = 15V$ at an angle suitable for the 1000-ohm load and select an operating point at about $V_{DS} = +10V$. Switch on the supplies and adjust V_{GS} to give this operating point.

Now apply an input of 2V peak-to-peak at 1000Hz from the generator and observe the output on the oscilloscope.

Question

4 Is the output a good sine-wave?

Measure the peak-to-peak output voltage and calculate the voltage gain $\dfrac{V_o}{V_i}$

Now insert the resistor as indicated in fig 11.9. Insert different values until you find one that makes the output signal about half its original size. This value is equal to the input resistance of the amplifier as fig 11.11 shows.

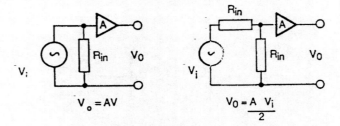

Fig 11.11 Input Resistance measurement

146

5 *The input resistance of fig 11.9 can not be greater than the bias resistor R_G. Is it, however, much less than this? If not, what does this indicate?*

PRACTICAL CONSIDERATIONS
AND APPLICATIONS

We have studied an N-channel JFET. A P-channel JFET is very similar in operation but uses reversed-bias voltages as in fig 11.12.

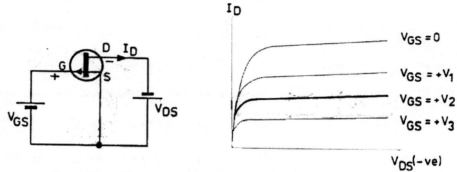

Fig 11.12 A P-channel JFET and Characterisitic

Another form of FET exists whose gate is insulated from the channel (insulated gate FET). The most common insulation method is a metal-oxide layer and the type is called **MOSFET**. The gate on a JFET must not be biased in such a way as to forward-bias the PN junction however, in a MOSFET, no such limitation applies.

It is therefore possible to bias the gate in either polarity. Fig 11.13 shows the usual characteristics of two types of N-Channel MOSFET.

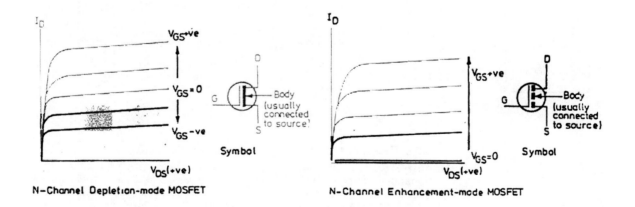

N-Channel Depletion-mode MOSFET

N-Channel Enhancement-mode MOSFET

Fig 11.13 N-channel MOSFET Characteristics and Graphical Symbols

The depletion-mode type is like a JFET with an extended gate bias range but the enhancement-mode is quite different since at $V_{GS} = 0$, no current flows at all. This is often a useful feature.

Two matching types of MOSFET exist with P-channel and have reverse polarities of bias. Their symbols are shown in fig 11.14.

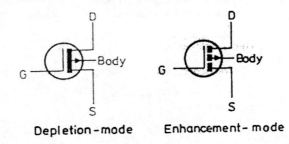

Depletion-mode   Enhancement-mode

Fig 11.14 P-channel MOSFET Symbols

We can put all these types together in a family tree as in fig 11.15.

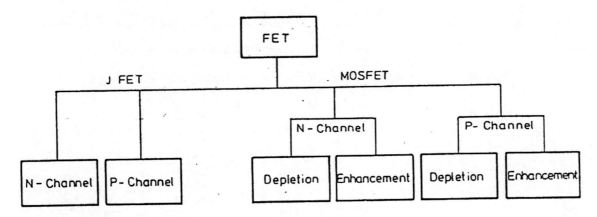

Fig 11.15 An FET Family Tree.

MOSFETs have even higher input impedances than JFETs

All FETs are useful for amplification with minimum load on the source. Enhancement-type MOSFETs are specially valuable as electronic switches because with no bias they are normally non-conducting and the high gate resistance means that very little control current is needed. MOSFETs, due to their very high gate resistance, can easily accumulate large static charges and can become damaged unless carefully handled. Some types are fitted internally with protective zener diodes to prevent damage during handling.

148

SUMMARY

In this assignment you have learnt that:

1. FETs can take various forms.

2. All forms possess high gate resistance, but particularly the MOS types.

3. As voltage amplifiers in a common-source circuit, FETs are not very linear (i.e they produce distortion if the output voltage is large).

4. FETs can be used as electronic switches.

EXERCISE

An important parameter of an FET used as an amplifier is its 'transconductance'. This is defined by:

$$\text{Transconductance } (g_S) = \frac{(\text{change in } I_d)}{(\text{change in } V_{GS})} \text{ mA}/\text{V}$$

(common source)

Study your graph, fig 11.7 and estimate the change in I_d for a 0.5V change in V_{GS} when $V_{DS} = 10V$ and $V_{GS} = 1.0V$. Then find g_s.

The voltage gain (A) for a load resistor R is given by

$$A = \frac{g_S R}{10^3} \text{ where R is in ohms } (\Omega)$$

Use this expression to verify the voltage gain $\frac{V_o}{V_i}$ measured in the Assignment.

$V_{GS}(V)$	I_D(mA) for $V_{DS} =$ (V)						
	0	0.5	1	2	5	10	15
0 −0.5 −1.0 −1.5 −2.0 −2.5							

Fig 11.6 JFET measurements

149

SECTION B: ━━━━━━━━━━━

BJT AND JFET AMPLIFIERS

OBJECTIVES

1. An understanding of a.c voltage and current gain together with input impedance in the bipolar transistor.

2. An understanding of a.c voltage gain in the field effect transistor.

EQUIPMENT REQUIRED

Qty	Apparatus
1	Electricity & Electronics Constructor EEC470
1	Amplifier Kit EEC473
1	Power supply unit 0 to +20V variable d.c.and ±15V d.c. regulated (e.g Feedback Power Supply PS445)
1	Function generator. Sinusoidal 8V pk to pk @ 1kHz (eg. Feedback FG601)
1	2–Channel Oscilloscope
1	Multimeter or
1	Voltmeter 25V dc

KNOWLEDGE LEVEL

Before working this assignment you should :

● Know the operation of circuits handling combined ac and dc voltages and currents.

● Know the meaning of the term 'dc bias '

150

INTRODUCTION

In Assignments 2 and 3 we saw how the bipolar transistor is able to amplify steady d.c signals applied to the base. In this assignment we will learn how a transistor amplifier can amplify a.c signals. An a.c signal is shown in fig 4.1a. As you can see the signal changes polarity every half cycle so it is necessary to supply the transistor with a d.c bias signal so that it amplifies over the entire cycle. A test circuit is shown in fig 4.1b. The varying collector current will cause a voltage drop across the load resistor R_L which will be opposite in polarity to the input signal. The output is taken from the capacitor. Fig 4.1c shows the voltages at important points in the circuit.

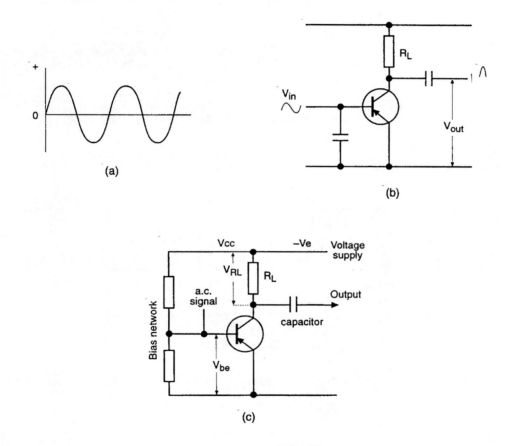

Fig 4.1 Behaviour of an a.c signal in an
NPN bipolar transistor amplifier

151

Amplifier Classification

The type of amplifier we have just described amplifies the entire cycle of the a.c input signal. It is called a Class A amplifier. In some applications we only require the positive half of the cycle to be amplified. A d.c bias voltage is not required in this case. It is then called a Class B amplifier.

A.C Gain

1. Voltage Gain:

This can be found by measuring the ratio of the peak-to-peak collector-emitter and base-emitter signal voltages.

$$\text{Voltage Gain} = \frac{\text{Collector–Emitter}}{\text{Base–Emitter}} \text{ peak–to–peak voltages}$$

2. Current Gain:

This can be found by measuring the ratio of the peak-to-peak collector and the base currents. The base signal current can be found by measuring the signal voltage drop across a known resistor as in fig 4.2.

$$\text{Current Gain } h_{fe} = \frac{\text{Collector}}{\text{Base}} \text{ peak-to–peak signal currents}$$

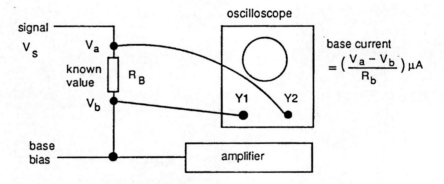

Fig 4.2 Circuit to measure base current

Input Impedance

The input impedance is the a.c load that a transistor imposes on the previous stage. It can be found by measuring the ratio of the peak-to-peak base-emitter voltage and the peak-to-peak base current.

$$\text{Input Impedance (Ohms)} = \frac{\text{Base–Emitter peak-to peak voltage [V]}}{\text{peak-to-peak Base Current [A]}}$$

Do not forget to use the right units for all values.

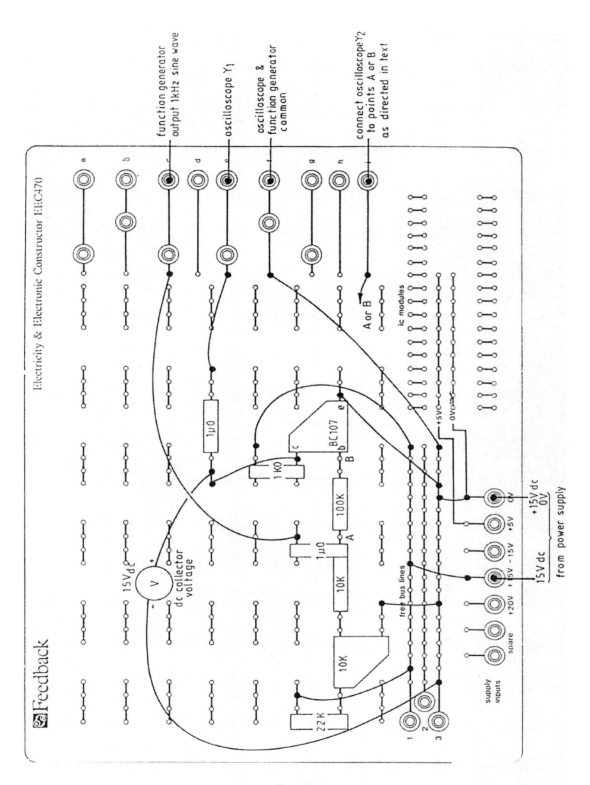

Fig 4.4

EXPERIMENTAL PROCEDURE

AC Gain in the bipolar transistor

First we shall measure the a.c current and voltage gain of a bipolar transistor amplifier.

The test circuit is shown in fig 4.3. Set up the layout in fig 4.4 and then check your layout with the circuit in fig 4.3. Switch on the power supply and function generator. Set the function generator output to zero, and adjust the base bias potentiometer for a d.c collector voltage of 8 volts (approximately half the supply voltage). Switch on the oscilloscope and display both Y1 and Y2. Turn up the function generator output until the output displayed on Y1 is 8 volts peak-to-peak. It should be sinusoidal.

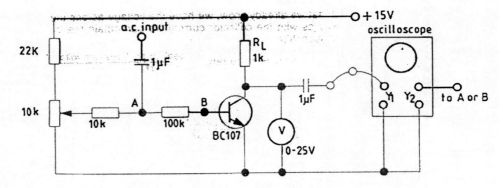

Fig 4.3 Test circuit for a.c current and voltage gain

Question

1   If the output signal was flat at the top or bottom of the sine-wave, what might the problem be

Measuring the Voltage Gain

We know the peak-to-peak collector voltage, so we need to measure the peak-to-peak base voltage. Connect the input of Y2 on the oscilloscope to point B on the circuit.

Measure the display on Y2 — it will be quite small.

From the relationship

$$\text{Voltage Gain} = \frac{\text{peak-to-peak collector signal voltage}}{\text{peak-to-peak base signal voltage}}$$

You can calculate the voltage gain.

Question

2   What is the voltage gain of the stage?

Measuring the Current Gain

We know the peak-to-peak collector signal voltage and we know the value of the collector load resistor (1kΩ). From Ohms law we can calculate the peak-to-peak collector current.

$$\text{peak-to-peak collector signal current} = \frac{\text{peak-to-peak collector resistor voltage}}{\text{collector load resistor}}$$

Now we need to know the peak-to-peak signal base current. We can calculate this by measuring the signal voltage drop across the resistor in series with the base (resistor A-B).

Measure the signal voltage at point A. By subtracting the signal voltage at point B that we already know, we have the voltage across the resistor. In the same way as with the collector current we can obtain the base current. From the relationship:

$$\text{peak-to-peak base current} = \frac{\text{peak-to-peak resistor voltage}}{\text{resistor value}}$$

We can now calculate the a.c current gain

$$\text{current gain} = \frac{\text{peak-to-peak collector signal current}}{\text{peak-to-peak base signal current}}$$

Questions

3 *What is the a.c current gain?*

4 *Are the values of a.c current and voltage gain dependent on resistor values?*

Measurement of the Input Impedance

We have the measurements required to calculate the input impedance of the bipolar transistor operating as an amplifier.

From the relationship:

$$\text{Input impedance (Ohms)} = \frac{\text{Signal peak-to-peak base voltage}}{\text{Signal peak-to-peak base current}}$$

We can calculate the a.c input impedance.

Question

5 *What is the a.c input impedance?*

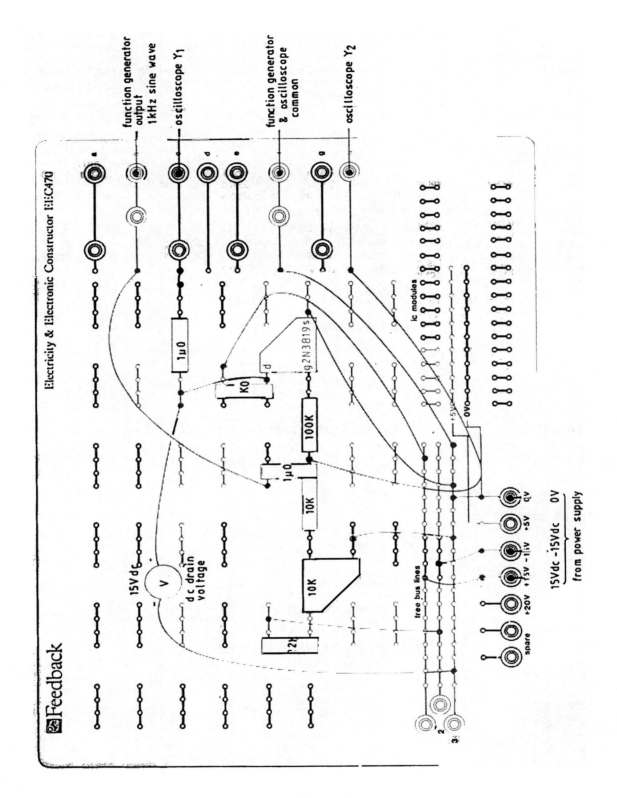

Figure 4.6

A.C Gain in the Field Effect
Transistor Amplifier

The junction field effect transistor can also be used as an a.c amplifier in a similar way to the bipolar transistor. The bias applied this time to the gate must be negative as we found in Assignment 3.

The test circuit we are going to use to measure the a.c voltage gain is shown in fig 4.5. As we found in Assignment 3 the gate draws almost no current, so we cannot measure the current gain.

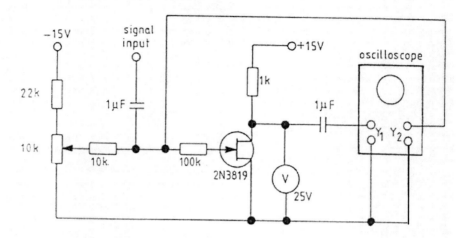

Fig 4.5 Test circuit for FET a.c voltage gain

Measuring the
Voltage Gain

Set up the layout in fig 4.6 and check your layout with the circuit diagram in fig 4.5. Switch on the power supply and adjust the bias potentiometer for a drain voltage of 10 volts on the meter. Switch on the function generator and the oscilloscope. Turn up the signal output of the function generator until you have a peak-to-peak signal voltage of 5 volts on the drain, as displayed on Y1. Measure the peak-to-peak voltage required on the gate as displayed on Y2.

From the ratio:

$$\text{Voltage gain} = \frac{\text{peak-to-peak signal drain voltage}}{\text{peak-to-peak gate voltage}}$$

you can calculate the a.c voltage gain of the amplifier.

Question

6   *What is the a.c voltage gain?*

7   *Does it depend on any resistor values?*

8   *What do you think is the input impedance of the field effect transistor in this amplifier?*

PRACTICAL CONSIDERATIONS AND APPLICATIONS

The application of amplifiers to a.c amplification is very useful. Many of the devices that we use every day make use of them. The telephone system, radio, television and industrial controls are examples.

Both the bipolar and field effect transistor are used in different applications, determined by the requirements for input impedance, voltage and current gain.

Testing for Malfunctions

By using an oscilloscope the signal voltages on the input can be examined and with a test meter the d.c levels can be checked. If the fault is in the amplifier there will be a signal at the input but not at the output. If the bias is correct, but there is the wrong d.c output level there is probably a transistor malfunction.

Using a function generator, like the FG601, a test signal can be injected into the amplifier for further tests. Input and output levels for both d.c and signal are usually given in the service manual.

SUMMARY

In this assignment you have learnt that

1 In order to amplify all the signal, d.c bias is required.

2 The bipolar transistor input impedance is low, so it operates on current. It has high current and voltage gain.

3 The FET input impedance is high, so it operates on voltage. The voltage gain is low but the current gain is very high.

EXERCISE

The transistor in fig 4.7(b) has a current gain of 250 while the FET of fig 4.7(a) has a voltage gain of 5. A current generator supplies a sinusoidal current of 0.5µA rms to be amplified. Assuming that suitable bias arrangements exist so that both amplifiers operate in class A, which circuit gives the greater a.c output voltage?

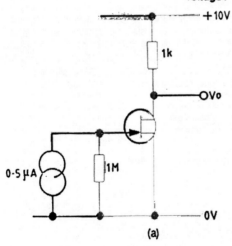

(a)

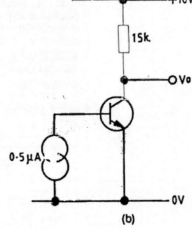

(b)

Fig 4.7

SECTION C:

FREQUENCY RESPONSE AND NEGATIVE FEEDBACK NETWORK

FREQUENCY RESPONSE

OBJECTIVES

1. To understand the meaning of frequency response, gain and phase shift in an amplifier.
2. To understand the effect of negative feedback on frequency response, gain and phase shift.

EQUIPMENT REQUIRED

Qty	Apparatus
1	Electricity & Electronics Constructor EEC470
1	Amplifier Kit EEC473
1	Power supply unit −15V d.c. regulated (e.g Feedback Power Supply PS445)
1	Function generator Sinusoidal 5V pk to pk @ 100kHz (eg. Feedback FG601)
1	2−Channel oscilloscope
1	Multimeter or
1	Voltmeter 15V dc

KNOWLEDGE LEVEL

See prerequisite requirements

INTRODUCTION
Bandwidth

In any system transmitting electrical signals, the transmission tends to fall off at high frequencies. There are various reasons for this. In the practical experiment you will do, the transistor's performance is the main factor causing the effect. A simpler and very common cause of it is shown in fig 8.1. Here a constant-current source is feeding the load resistance R_L. Stray capacitance provides a path, whose reactance we may call X_C, through which some of the signal current is diverted .

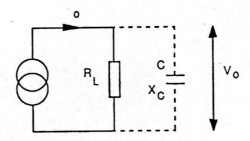

Fig 8.1

Since $X_C = \dfrac{1}{2\pi fc}$ it will reduce as the frequency increases, effectively reducing the total load impedance Z_o.

As $V_o = i_o Z_o$ for a given output current i_o the output voltage V_o will reduce as the frequency increases.

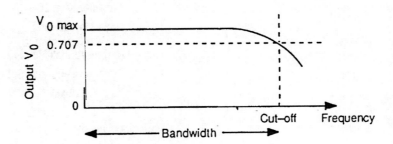

Fig 8.2 Frequency response of amplifier

A graph of output against frequency can then be drawn as in fig 8.2. The useful frequency range or bandwidth is that frequency at which the output power is down to half. Now power is V_o^2/R_L. The load R_L is constant so the half power frequency or cut-off frequency occurs when Vo^2 is half its low frequency value Vo^2 (max).

Thus $V_o^2 = \dfrac{V_o^2 (max)}{2} = .707 V_o (max$

Phase change

The effect of the amplifier reactance shown in fig 8.1 is to cause a phase change between signal and output. This phase change will increase with frequency. At the cut-off frequency the phase change is 45°.

Effect of negative feedback on frequency response

In Assignment 7 we saw that in the negative feedback amplifier fig 8.3 a fraction ß of the output signal V_o is subtracted from the input V_{in}. It is this difference that is applied to the input of the amplifier.

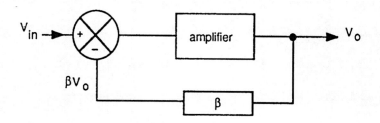

Fig 8.3 Negative feedback amplifier

The relation of the input V_{in} to the output V_o is

$$\frac{V_o}{V_{in}} = \text{overall gain} = \frac{A}{1 + ßA} \cdot$$

(where A is the gain of the amplifier without feedback).

*Proof of this equation is given in Appendix A.

If A is large then ßA is much greater than 1 so that the overall gain is roughly equal to $\frac{1}{ß}$

Now if A falls with increasing frequency as in fig 8.2, provided ßA is still large, the overall gain does not change much. So negative feedback tends to reduce gain change. This will also increase the frequency at which cut-off occurs.

Effect of negative feedback on phase lag

The same reasoning can be applied to the effect of negative feedback on phase lag as applied to gain. For a given frequency the phase lag with negative feedback will be less than without it.

General effect of negative feedback

If the output changes for any reason, the negative feedback affects the input in such a way as to counteract the change.

161

EXPERIMENTAL PROCEDURE

Frequency Response

In this activity we shall measure the gain of a transistor amplifier at various frequencies comparing the response with and without negative feedback. The transistor used is the PNP AC128, and the feedback introduced by adding a resistor in the emitter circuit.

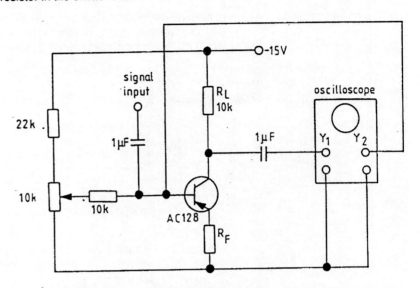

Fig 8.4 Test circuit for frequency response

The test circuit is shown in fig 8.4.

Set up the layout in fig 8.5 and check it against the test circuit in fig 8.4.

Make R_F = 0 by inserting a link.

Set the frequency to 1kHz.

With the function generator set to zero, set the d.c collector voltage to 8 volts. Turn up the function generator output until there is an 8-volt peak-to-peak signal displayed on Y1. Monitor the input on Y2. It will be quite small.

Measure the voltage gain:

$$\frac{\text{peak-to-peak collector voltage}}{\text{peak-to-peak base voltage}}$$

Copy the results table as shown in fig 8.6, reproduced at the end of this assignment, and enter the value.

Measure the gain at the other frequencies in the table entering the results in the gain column.

Notice the gain reduction at high frequencies.

162

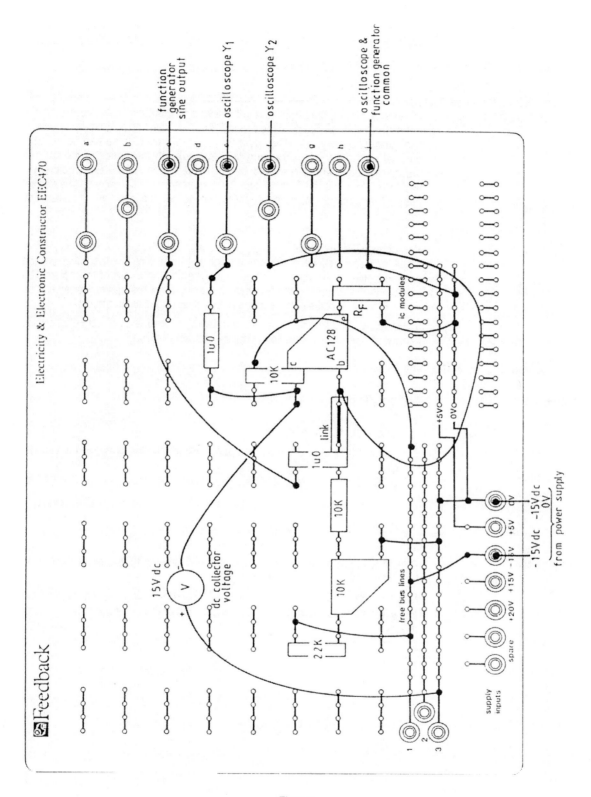

Fig 8.5

PRACTICAL CONSIDERATIONS AND APPLICATIONS

In the experiment, the negative feedback was proportional to the current in the transistor. This feedback tended to stabilise that current against the effects of changing frequency. Note however that if extra capacitance had been put in parallel with the load, worsening the drop in high-frequency response, the feedback signal would not have been affected, and so the feedback would have no effect on the loss of high frequencies from this cause. (To correct this, a feedback signal proportional to the output voltage would have been required)

Negative feedback plays an important part in the design and stabilisation of amplifier circuits. Where noise is introduced internally the feedback can be applied between the output and point just before the origin of the noise. The noise is then reduced by the feedback but the gain of the earlier stages can be increased. Thus noise can be reduced by using negative feedback.

It also allows amplifiers to be produced having a fixed level of gain over a wide bandwidth and offsets variation in device characteristics.

SUMMARY

In this assignment you have learnt that:

1   The bandwidth of an amplifier is the frequency range in which the gain is within 70.7% of the maximum.

2   At cut-off the phase change compared with the phase change at maximum gain is 45°.

3   By using negative feedback the bandwidth can be extended.

4   By using negative feedback the phase change occurring with frequency is reduced.

EXERCISE

Refer back to Assignment 7 and calculate the value of ß, the feedback fraction, for the amplifier in this assignment.

Now choose a frequency in the results table and by taking the voltage gain with and without feedback, calculate the practical value of ß. See if it agrees with the theory.

Frequency	Voltage gain		Phase change	
	$R_F = 0$	$R_F = 220\Omega$	$R_F = 0$	$R_F = 220\Omega$
1kHz				
5kHz				
10kHz				
30kHz				
60kHz				
100kHz				

Fig 8.6.

LAB 7

Operational Amplifiers

166

SECTION A:

INVERTING OPERATIONAL AMPLIFIER

THE OPERATIONAL AMPLIFIER

OBJECTIVES

1. A knowledge of its differential input and high gain.
2. An understanding of its use as a summing amplifier.

EQUIPMENT REQUIRED

Qty	Apparatus
1	Electricity & Electronics Constructor EEC470
1	Amplifier Kit EEC473
1	Power supply unit 0 to +20V variable d.c and ±15V d.c. regulated (e.g Feedback Power Supply PS445)
2	Multimeters or
2	.Voltmeters 15V dc

PREREQUISITE ASSIGNMENTS

KNOWLEDGE LEVEL

Before working this assignment you should :

● Know the operation of ac coupled amplifier circuits with and without negative feedback applied.

INTRODUCTION

In the amplifiers we have looked at so far the inputs and outputs have been a.c coupled. This means that they are of no use for very low frequencies or for d.c amplification. The main problem in making an amplifier for d.c is in ensuring that the mean d.c level of the output is the same as the mean d.c level at the input. In the single stage in the previous assignments this is obviously not so.

Operational Amplifiers

The operational amplifier is a device that has all the properties required for d.c amplification. It contains several stages and circuitry for temperature drift compensation. The gain of the amplifier is far in excess of that available using single transistors. Although it could be made using discrete transistors it is usually an integrated circuit (IC) with all the components on a single silicon chip. This makes operational amplifiers available in large quantities at very low cost.

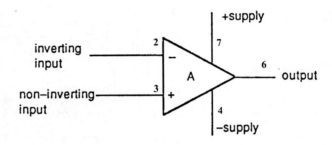

Fig 10.1 Operational Amplifier

Fig 10.1 shows the circuit symbol for an operational amplifier (often abbreviated to 'Op-amp').

We can see some important features:

1. It has positive and negative power connections. This is so that the output can swing either side of zero volts.

2. It has a positive and a negative input. This means that signals applied to the positive input are not inverted while signals applied to the negative input are inverted. Both inputs have the same total gain. This is called a differential input stage.

3. Although it is not shown on the symbol, the amplifier gain is very high. 100,000 would be a typical value.

In a previous assignment we saw the benefits of negative feedback. We found that by sacrificing some gain the stage gain can be made independent of the device characteristics and be determined by circuit components that can be controlled very closely. As the gain in the op-amp is so high we can afford to lose some by using negative feedback thus producing a very stable amplifier system. We could feed some of the output back to the input — but to make the feedback negative we must feed it back to the negative input. The negative input will now have a different signal on it than the actual input voltage (the sum of the input and the negative feedback). We must separate the real input from the inverting input by a resistor.

The amplifier is designed so that when both inputs have zero volts on them the output has zero volts on it. As we are not using the positive input we must connect it to ground. Combining these ideas we have the circuit in fig 10.2.

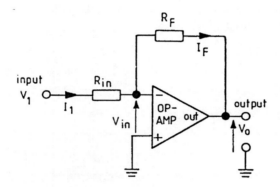

Fig 10.2 Op -amp feedback circuit

We will now apply some simple theory to this circuit.

We know that the gain of the op-amp is very high, so V_{in} will be very small compared with V_o. If we assume it to be almost zero we can say:

$$V_o = -I_F R_F \text{ (very nearly)} \quad \therefore \quad I_F = \frac{-V_o}{R_F}$$

$$\text{but } I_F = I_1 = \frac{V_1}{R_{in}}$$

$$\frac{V_o}{R_F} = \frac{-V_1}{R_{in}} \quad \therefore \quad \frac{V_o}{V_1} = \frac{-R_F}{R_{in}}$$

This is very important as we have produced an amplifier the gain of which is the ratio of two resistors, and is independent of the actual gain of the op-amp. This is called 'closed loop operation'. Notice that the amplifier inverts the signal.

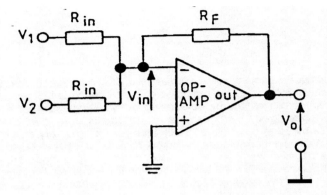

Fig 10.3 Summing amplifier

If we add another input resistor equal to the value of R_{in} as in fig 10.3, we can modify the above equation thus:

$$I_F = I_1 + I_2 = -\left(\frac{V_1}{R_{in}} + \frac{V_2}{R_{in}}\right)$$

$$V_o = -\left(\frac{R_F}{R_{in}}V_1 + \frac{R_F}{R_{in}}V_2\right) = -\frac{R_F}{R_{in}}(V_1 + V_2)$$

We now have an amplifier that produces an output proportional to the SUM of the two input voltages. This is called a summing amplifier.

If the two input resistors were not equal the equation would be modified thus

$$V_o = -V_1\left(\frac{R_F}{R_{in1}}\right) - V_2\left(\frac{R_F}{R_{in2}}\right)$$

The operational amplifier we shall be using for our experiments is a type 741. It is very small, and comes in a plastic pack 9mm x 6mm x 3mm. For convenience the device is soldered on a small board with pins to suit the EEC470 deck. There are a few practical problems. Although in theory when both inputs are zero the output should be zero, due to leakage currents and other causes there is a slight 'offset' of the output. This can be removed by the offset null potentiometer mounted in the module.

In the first activity we shall familiarise ourselves with the IC and try the offset null control. In the second activity we shall make a closed-loop feedback circuit.

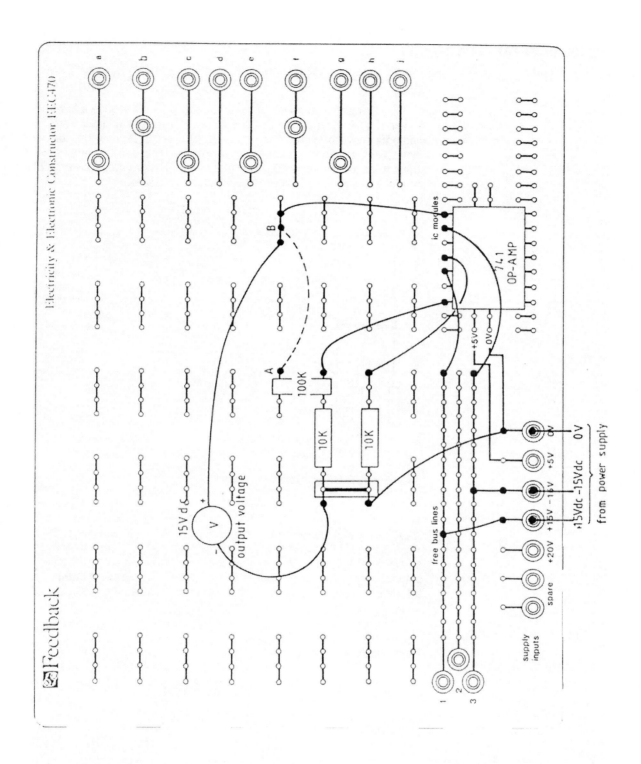

Fig 10.5

EXPERIMENTAL PROCEDURE

Dc Offset

The circuit we shall use is shown in fig 10.4. Set up the layout in fig 10.5.

NOTE

Operational amplifiers of the type used in the Kit require a fixed power supply of +15, 0, −15V dc. This fixed supply is usually omitted from circuit diagrams showing operational amplifiers, in order to reduce the number of connections shown to the minimum necessary. The +15V dc and −15V dc connections to pins 12 and 9 respectively of the 741 Op–Amp on fig 10.5 are therefore omitted from fig 10.4 and subsequent diagrams.

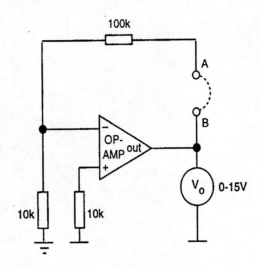

Fig 10.4 Test circuit for dc offset

Switch on the power supply. Turn the potentiometer, marked 'zero', in the op-amp module with a small screwdriver. Notice that the output changes between +15 volts and -15 volts very suddenly. It is not possible to set it to zero as the amplifier is operating in the open-loop mode with very high gain. The potentiometer is not fine enough in this condition. Now link A to B thus adding a 100kohm feedback resistor. The loop is now closed and the gain reduced.

Questions

Try to null the amplifier again.

1   Can you set the output to zero?   Why?

172

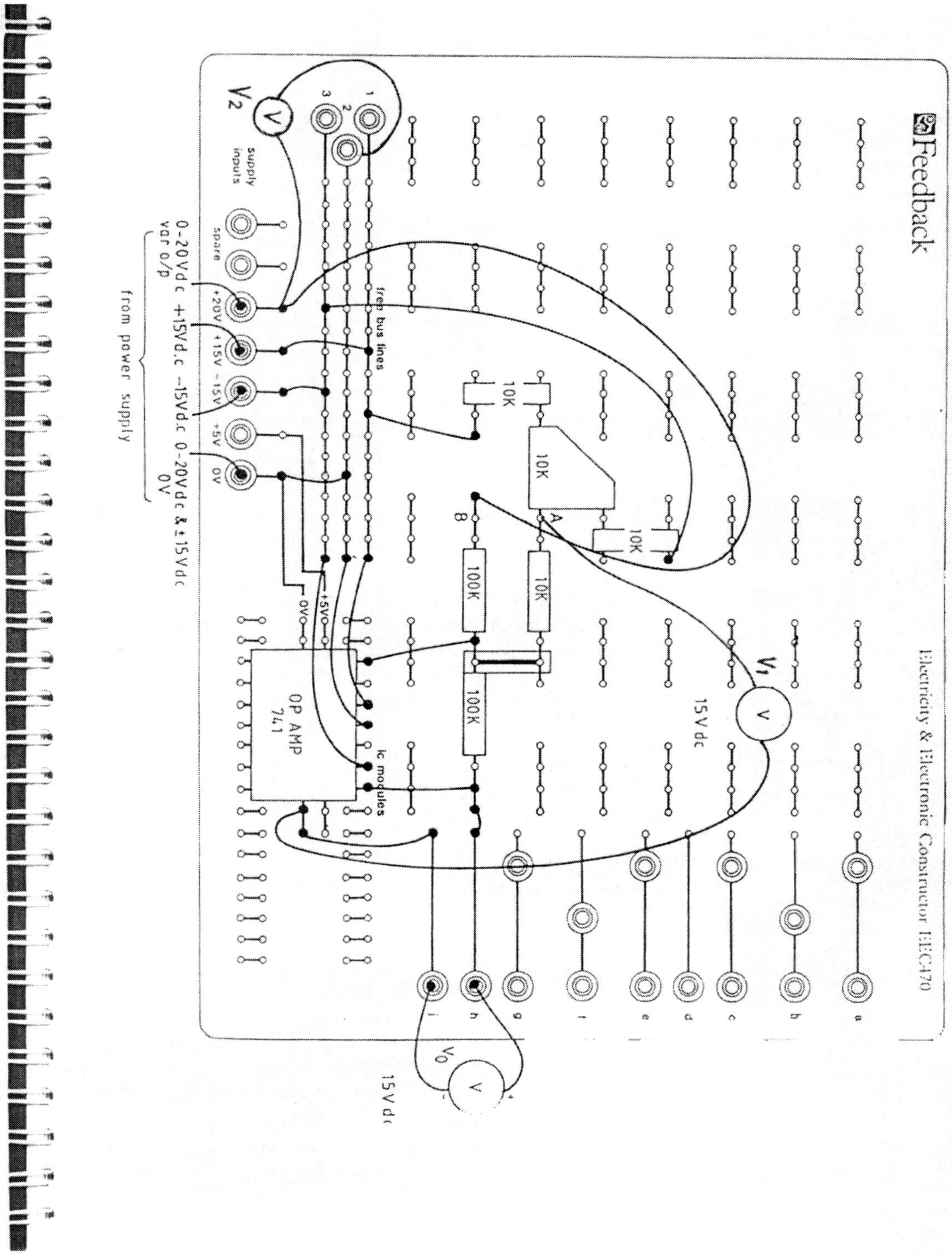

Feedback
Electricity & Electronic Constructor EEC470
supply inputs
spare
free bus lines
ic modules
OP AMP 741
10K
100K
15Vdc
0-20Vdc var o/p
+15Vdc
-15Vdc
0-20Vdc & ±15Vdc
0V
from power supply

Op Amp Performance

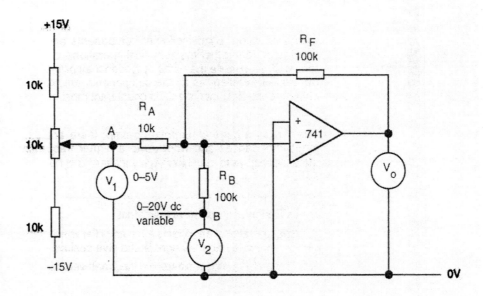

Fig 10.6 *Test circuit for Op Amp performance*

Set up the layout in fig 10.7. Check it against the circuit diagram in fig 10.6. The circuit is a summing amplifier. The variable voltage from the power supply is one input and the potentiometer is the other. Notice that R_A is different from R_B. The output is related to the input by the ratio of the feedback resistor to the input resistor for each input.

Question

2 What is the relationship of the output to the input voltage?

It will take the form:

$$- V_o = xV_1 + y\,V_2$$

where x and y are resistance ratios.

Turn on the equipment and turn the input voltage controls up and down. Notice the polarity change in the output voltage V_o

Copy the results table as shown in fig 10.8, reproduced at the end of this assignment. Set up each input voltage condition in the table. Record the output voltage for each condition. Using the formula you found in Q3, calculate the expected output voltage each time and enter it in the table.

We have made an electronic adding machine as the output voltage is the algebraic sum of the input voltages. It uses electric current as an analogue of the number. There are larger systems using this principle and they are called analogue computers.

174

PRACTICAL CONSIDERATIONS
AND APPLICATIONS

The operational amplifier has a wide variety of uses in control systems and instrumentation. By using other components around it the op-amp may be used to perform other mathematical operations, differentiation and integration, for example. (Hence the name operational amplifier). The thermal stability of an op-amp is excellent as all the components are mounted on the same silicon chip, and the designer is able to counteract most of the drift problems.

The 741 is a general-purpose op-amp; there are others which have higher input impedances and work up to a higher frequency. Although the 741 is quite insensitive to supply voltage variations it is normally used with a regulated supply.

SUMMARY

In this assignment you have learnt that :

1   The operational amplifier is a d.c amplifier which can amplify both positive and negative signals and give positive or negative outputs.

2   The amplifier has a differential input and very high gain.

3   When operated in a closed loop the gain can be very closely controlled by resistors.

4   It can sum two separate inputs.

$$Vo = - (xV_1 + yV_2)$$

Input voltage		Output voltage	
V1	V2	Vo	Calculated Voltage
0.5	2		
0.1	6		
0.3	4		
−0.9	2		
−1.1	4		
−1.5	6		

Fig 10.8

SECTION B:

NON-INVERTING OPERATIONAL AMPLIFIER

Non-Inverting Amplifier Circuit

Figure 10.9 shows the non-inverting amplifier circuit connection.

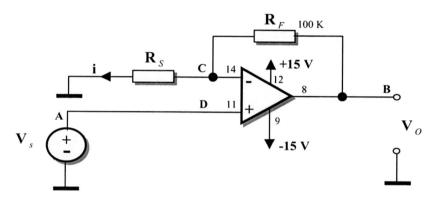

Figure 10.9

Because the input impedance of the OP amplifier is infinite big, so the input current of the OP amplifier can be considered as close to zero.

Based on the assumption above, the equation of the voltage can be listed as:

$$i = \frac{V_C}{R_1} = \frac{V_O - V_C}{R_F} \qquad \Rightarrow \qquad \frac{Vo}{V_S} = \frac{R_F + R_S}{R_S} = (1 + \frac{R_F}{R_S})$$

$$V_D \cong V_C \cong V_S$$

Question 1: Using R$_S$ = 1 K, design the circuits to have nominal voltage gains of 1, 11, 101 and 1001 as given for the ideal OP amplifier.

Question 2: Measure the circuit gains, and compare the measured values with the calculated values of the gain. Is there any error between them? If it is, where these errors come from?

177

SECTION C:

FREQUENCY RESPONSE OF OP AMPLIFIER

Frequency Response of Non-Inverting Amplifier Circuit

Using the circuit at Figure 10.9 in Section B above.

- o Connect the circuit for the voltage gain of 1 and measure the voltage gain at frequency of 100 Hz.
- o Connect the circuit for the voltage gains of 11, 101 and 1001 and measure the voltage gains again at 100 Hz.
- o Measure the gain of your circuit of 1001 at a frequency of 10 KHz.

Question 1: List the different gains and different frequency values measured above, what is the deal?

- o Connect the circuit for the voltage gain of 101, measure the voltage gains at frequency of 10 Hz, 100 Hz, 1 KHz, 10 KHz, 100 KHz and 1 MHz. Convert the gain to the dB unit and finish the graphic in Figure 10.10 (dB = 20 log(**gain**)).

Question 2: Based on the finished Figure 10.10, what is your conclusion for the amplifier circuit's gain with the frequency?

- o Connect the circuit for the voltage gain of 101, measure the voltage gains at a frequency in which the gain is decreased 3 dB compared with the gain in the low frequency, and record this gain and the frequency, A $_{o1}$ and f_{wd1}.
- o Connect the circuit for the voltage gain of 1001, measure the voltage gains at a frequency in which the gain is decreased 3 dB compared with the gain in the low frequency, and record this gain and the frequency, A $_{o2}$ and f_{wd2}.

Question 3: Based on the measurement above, calculate the products of A $_{o1}$ $\times f_{wd1}$ and A $_{o2}$ $\times f_{wd2}$. What is the relationship between them?

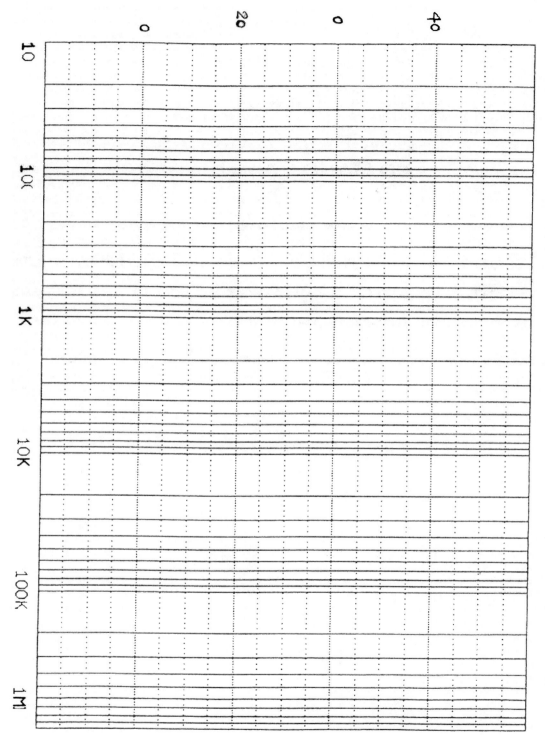

Figure 10.10

179

LAB 8

Low-Pass, High-Pass
and Band-Pass Filters

- ❑ SECTION A: IN-ACTIVE LOW-PASS AND HIGH-PASS FILTERS
- ❑ SECTION B: ACTIVE LOW-PASS AND HIGH-PASS FILTERS
- ❑ SECTION C: ACTIVE BAND-PASS FILTERS

SECTION A:

IN-ACTIVE LOW-PASS AND HIGH- PASS FILTERS

In-Active Filter: Composed of the pure resistors and capacitors. No any OP Amplifiers are used in the filter.

Cut-Off Frequency f_c: The frequency where the amplitude of the filter output is reduced by 3 dB (0.707 times of the normal amplitude of the output).

$$f_c = \frac{1}{2\pi RC} \tag{1}$$

where R is Ohms, and C is Farads, and the frequency is Hz.
Equation (1) is applied for both low-pass and high-pass filters.

Experiment steps:

- Construct inactive low-pass filter
- Calculate the cut-off frequency
- Connect the sinusoidal signal (from the functional generator) to the input of the low-pass filter
- Using the oscilloscope to monitor the output of the low-pass filter
- Varying the input signal's frequency from 0 – 500 KHz
- Record the output for each frequency with frequencies of 100 Hz, 1 KHz, 10 KHz, 100 KHz, 500 KHz
- Draw the output of the low-pass filter with frequency as the variable in x-axis and the amplitude of the output as the variable in y-axis

Perform the same sequence for the inactive high-pass filter.

Make connections as shown in Figure 8-1 (a) and (b) to get an inactive low-pass and an inactive high-pass filter.

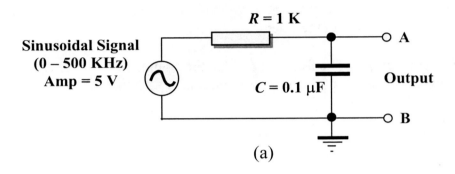

(a)

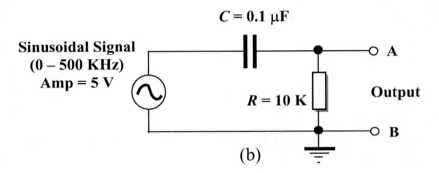

(b)

Figure 8-1.

Question 1: In which frequency, the amplitude of the output of the low-pass filter begin to drop to around a –3 dB?

Question 2: In which frequency, the amplitude of the output of the high-pass filter begin to get the normal value?

Question 3: Based on the questions above, why we call the first filter a low-pass filter, and the second filter a high-pass filter?

Question 4: Draw the frequency response for both low-pass and high-pass filter at the following graphs [Amplitude unit is 20log(Amp)].

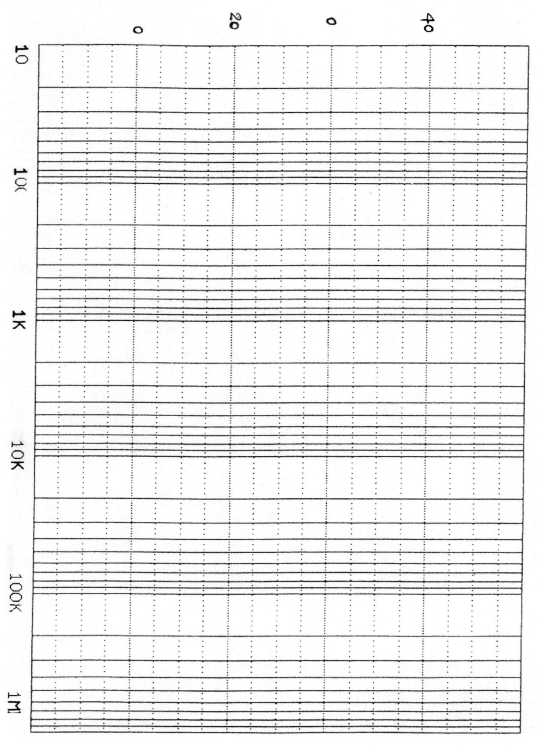

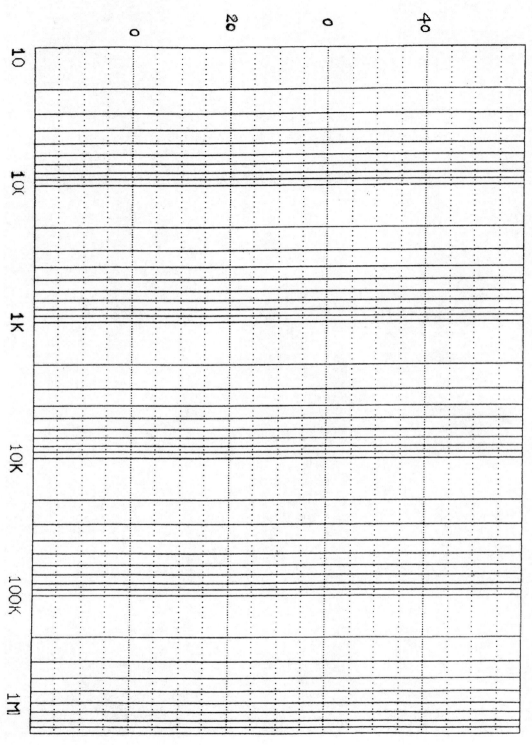

SECTION B:

ACTIVE LOW-PASS AND HIGH- PASS FILTERS

Active Filter: Composed of the combination of the OP integrated amplifier, resistors and capacitors. The OP Amplifier is considered as an active device in the filter.

Cut-Off Frequency fc: The frequency where the amplitude of the filter output is reduced by 3 dB (0.707 times of the normal amplitude of the output).

$$fc = \frac{1}{2\pi RC} \tag{1}$$

where R is Ohms, and C is Farads, and the frequency is Hz.
Equation (1) is applied for both low-pass and high-pass filters.

Experiment steps:

- Construct active low-pass filter
- Calculate the cut-off frequency
- Connect the sinusoidal signal (from the functional generator) to the input of the low-pass filter
- Using the oscilloscope to monitor the output of the low-pass filter
- Varying the input signal's frequency from 0 – 500 KHz
- Record the output for each frequency with frequencies of 100 Hz, 1 KHz, 10 KHz, 100 KHz, 500 KHz
- Draw the output of the low-pass filter with frequency as the variable in x-axis and the amplitude of the output as the variable in y-axis

Perform the same sequence for the active high-pass filter.

Make connections as shown in Figure 8-2 (a) and (b) to get an active low-pass and an active high-pass filter.

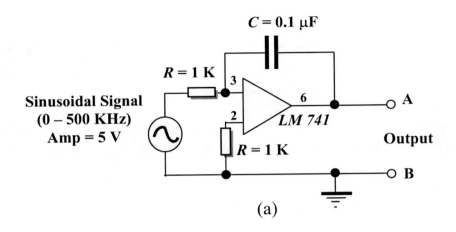

(a)

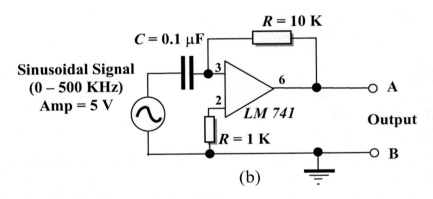

(b)

Figure 8-2.

Question 5: In which frequency, the amplitude of the output of the low-pass filter begin to drop to around a −3 dB?

Question 6: In which frequency, the amplitude of the output of the high-pass filter begin to get the normal value?

Question 7: Based on the questions above, why we call the first filter a low-pass filter, and the second filter a high-pass filter?

Question 8: Draw the frequency response for both low-pass and high-pass filter at the following graphs [Amplitude unit is 20log(Amp)].

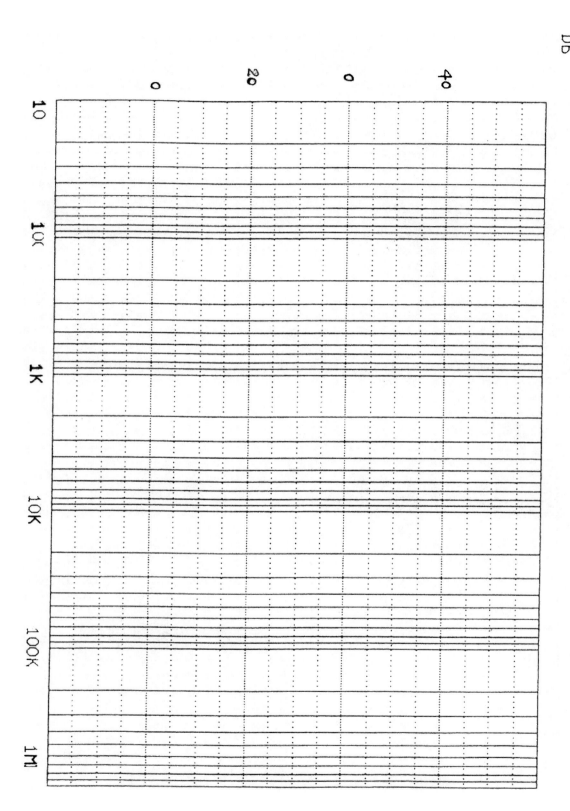

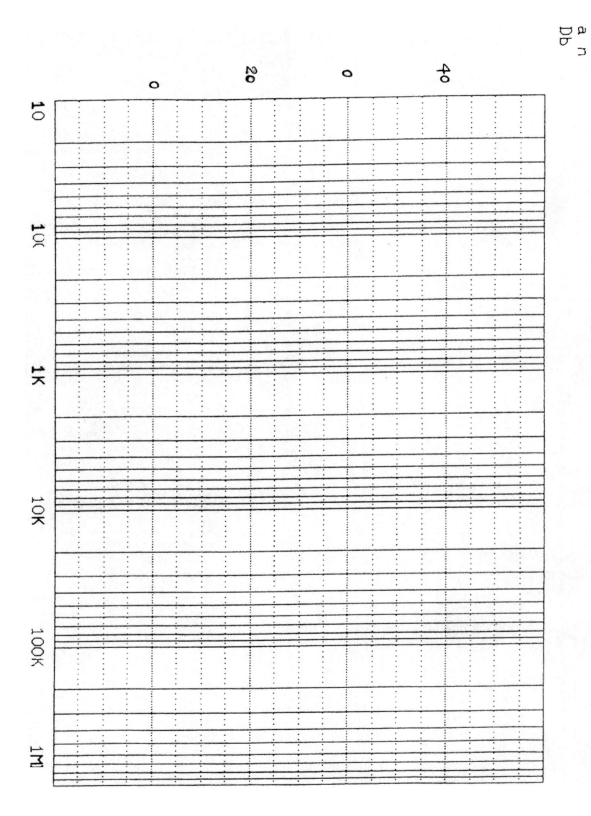

Question 9: Compared with the inactive low-pass and high-pass filters, what are the advantages of the active low-pass and high-pass filters?

Question 10: Can we combine a low-pass filter with a high-pass filter to obtain a band-pass filter? If it is yes, draw a schematic for this combination.

SECTION C:

ACTIVE BAND-PASS FILTERS

Cut-Off Frequency fc: The frequency where the amplitude of the filter output is reduced by 3 dB (0.707 times of the normal amplitude of the output).

$$fc = \frac{1}{2\pi RC} \qquad\qquad (1)$$

where R is Ohms, and C is Farads, and the frequency is Hz.
Equation (1) is applied for both low-pass and high-pass filters.
For a band-pass filter, we have two cut-off frequencies.

Experiment steps:

- The cut-off frequency for low-pass filter is around 1600 Hz, and the cut-off frequency for high-pass filter is around 150 Hz
- Based on equation (1), C_L and C_H values shown in Figure 8-3 below, and cut-off frequencies provided above, calculate the R_L and R_H values (around 1 KΩ or 10 KΩ)?
- Construct an active band-pass filter (refer to Figure 8-3)
- Connect the sinusoidal signal (from the functional generator) to the input of the band-pass filter
- Using the oscilloscope to monitor the output of the band-pass filter
- Varying the input signal's frequency from 0 – 500 KHz
- Record the output for each frequency with frequencies of 100 Hz, 200 Hz, 1 KHz, 1500 Hz, 2000 Hz, 10 KHz, 100 KHz, 500 KHz
- Draw the output of the band-pass filter with frequency as the variable in x-axis and the amplitude of the output as the variable in y-axis

You can collect more amplitudes of the output of the band-pass filter around two cut-off frequencies.

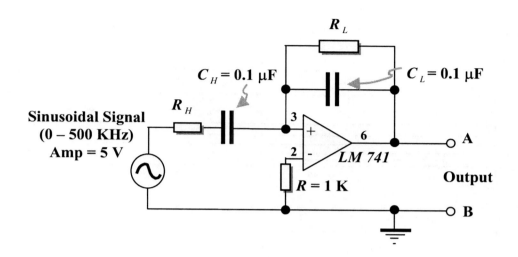

Figure 8-3.

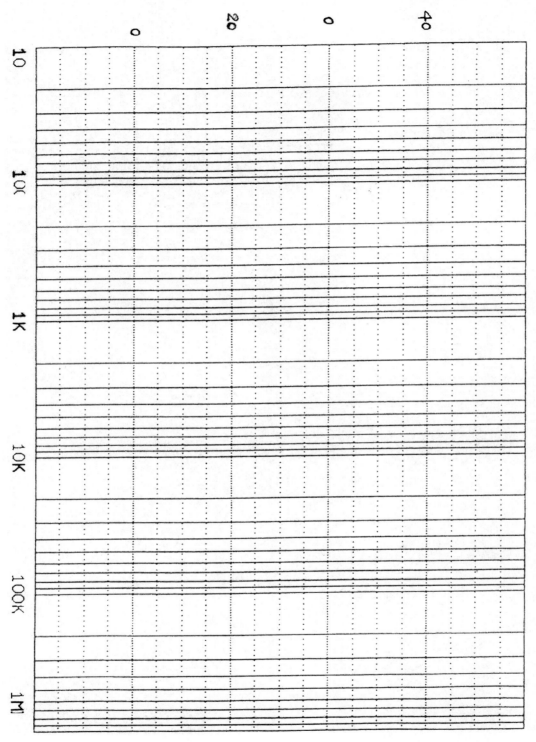

Question 11: Does the band-pass filter allow only a part of input signal pass through? If it is, what is the range of the passed signal's frequencies?

Question 12: Can we design an inactive band-pass filter? If it is, please draw a schematic of that kind of band-pass filter.

LAB 9
ANALOG-TO-DIGITAL CONVERTER

☐ **ANALOG-TO-DIGITAL CONVERTER**

Objective: Understand the operation principle of the A-D converter. Design and implement an 8-bit Analog-to-Digital circuit using ADC0804. Familiar the different operation modes of the ADC.

Equipment: ADC0804, 7-segment LEDs, resistors and capacitors.

INTRODUCTION

In this experiment you will become familiar with an analog-to- digital converter (ADC) integrated circuit. The ADC is used to convert an analog signal to a digital representation. Once the signal is represented digitally, it can be processed by a computer in any number of interesting ways. The analog signal might be an audio signal or it may be a video signal or it may be the output of a thermometer or some other sensor-based system.

DIGITAL REPRESENTATION OF ANALOG SIGNALS

Analog signals are continuously variable over an interval of values, such as 0-10 volts, -15 to +40 *mA*, etc. Digital approximations of exact values can be made by breaking the continuously variable interval into discreet intervals, as shown in Figure 1, where a 0-10 volt continuously variable interval is shown broken into 10 parts. The electronic process is very similar to the mechanical process of measurement of distance with a ruler that has been broken down into fractions of inches or meters.

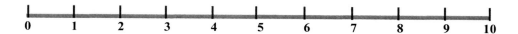

Figure 1. Breaking a Continuous Interval into Discreet Parts

In Figure 1, then, the rule for converting an analog quantity into a digital quantity might be to assign the analog quantity to the next highest digital quantity. Thus, analog 7.5 would become digital 8 and analog 5.05 would become digital 6. Obviously, an error of 10 per cent is possible with this representation scheme.

To achieve greater accuracy, it is necessary to break the continuous interval into a larger number of discreet parts. Breaking it into 100 parts limits the error to 1 per cent and breaking the interval into 1000 parts limits the error to 0.1 per cent.

The nature of digital electronic circuits makes the binary number system a natural choice for representation of discreet quantities, since it has only two digits, 0 and 1. For any number system, the number of distinct digits is equal to the system base. For example, base 10 has the digits 0-9, while base 8 has the digits 0-7. Numbers are simply the sum of the 1s, 10s, 100s, etc. columns, or the 1s, 2s, 4s, 8s, etc. columns. Figure 2 shows a comparison of base 10 and base 2 number representations.

195

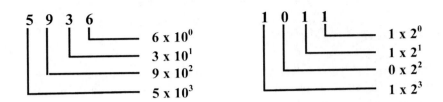

Figure 2. Base 10 (decimal) and Base 2 (binary) Number Representations.

Note also that Figure 2 indicates how to convert binary numbers to decimal numbers. Binary 1011 is decimal 11. Decimal units are referred to as digits, while binary units are referred to as bits. The decimal number 5936 is a four-digit number. The binary number 1011 is a four-bit number. An 8-bit binary number is called a "byte" The 2^0 bit is called the Least Significant Bit (LSB) and the last bit to the left is called the Most Significant Bit (MSB) . The LSB represents the smallest portion of the interval.

Note now that if an interval is broken into 10 parts, and if it is desired to use a binary representation of these parts, it will be necessary to use 4-bit binary numbers to uniquely represent each of the 10 parts. One possible representation becomes:

Decimal   0   1   2   3   4   5   6   7   8   9   10
Binary   0000 0001 0010 0011 0100 0101 0110 0111 1000 1001 1010

Note, however, that 4-bit binary numbers can represent decimal numbers from 0 to 15. Thus, the binary representations from 11 to 15 are not used. What this means is that with the same 4-bit representation it would have been possible to break the interval into 15 parts rather than just 10 parts.

Each additional bit doubles the number of decimal numbers which can be represented, thus doubling the accuracy of the representation. Five bits enables counting to 31, six enables counting to 63, etc. Thus, to divide an interval into 100 parts, 7 bits of binary representation will be needed. An 8-bit representation enables breaking an interval into 255 parts. How many bits would be needed to produce an accuracy of 0.1%?

BINARY CODED DECIMAL REPRESENTATION

Usually after a binary representation of a number is processed digitally, it is desirable to present the result as a decimal result, since most people are not accustomed to dealing with binary numbers. Sometimes it is convenient from a circuitry standpoint to represent each decade of a decimal number with a 4-bit binary number. This representation is called Binary Coded Decimal, or BCD. For example, the binary and BCD representations of 185 are

     1   0   1   1   1   0   0   1   (Binary)   0001   1000   0101   (BCD)

Notice that it only takes 8 bits in binary to represent 185, but it takes 12 bits to represent 185 in BCD. So the price paid for the circuitry convenience is the wasting of some bits. It is often possible to select between binary and BCD outputs when choosing certain integrated circuits.

THE ANALOG-TO-DIGITAL CONVERTER

The Analog-to-Digital Converter (ADC) converts an analog input signal to a digital output signal in either binary or BCD representation, depending on the choice of ADCS. Precision and speed are the two main considerations in the selection of an ADC. Both affect the cost of the selected unit.

The precision of an ADC is the number of bits into which it breaks down the analog interval. Perhaps the most common ADCs are 8-bit units, which are relatively inexpensive ways to achieve 0.5 per cent precision (1 part in 255). For greater precision at higher price, it is possible to obtain up to 22-bit ADCS. (How precise is 22 bits?)

It is important not to confuse accuracy with precision. It is possible to have a 10-bit unit, which should be able to accurately break an interval into 1027 parts, which has an accuracy of only 1 per cent rather than 0.1 per cent. Usually, however, manufacturers do not bother to achieve the precision unless they can also achieve the accuracy. Sometimes, however, the accuracy depends on externally supplied voltages.

The speed of an ADC is the speed which it takes to complete a conversion. A SERIAL ADC converts an analog signal one bit at a time and is thus slower than a PARALLEL ADC, which converts all bits simultaneously. A really fast ADC is called a FLASH ADC. Again, the simple rule is "the faster, the more costly." Thus, an 8-bit serial ADC should be cheap and a 13-bit flash ADC can be expected to be expensive. A nice compromise is achieved with the successive approximation ADC, which operates at an intermediate speed.

If the quantity to be converted is a d-c quantity, then conversion speed is not of concern, and a serial ADC is adequate. If the quantity to be converted is an audio signal, then conversions must take place approximately 50,000 times per second, and serial conversion is not quite good enough. A parallel ADC would probably be used. Guess what kind of ADC will be used in this experiment.

THE NATIONAL SEMICONDUCTOR ADC0804LCN ADC

The NS ADC0804LCN (0804) is an 8-bit, CMOS, successive approximation A/D converter which uses a differential potentiometric ladder. It has differential analog inputs which allow for high CMRR and compensating for any analog offset voltage. Its analog input range is recommended to be 5.0 volts, but can be reduced to a smaller interval to decrease the least significant bit of resolution. Its digital output is binary. Figure 3 shows the ADC pin connections.

The 0804 depends on a clock signal for coordination of the conversion process. A clock is simply a square wave generator, and the 0804 can be connected to provide its own clock signal or to use an externally applied clock signal. The resistor and capacitor connected at pins 19 and 4 are the resistance and capacitance fed back from output to input of an astable multivibrator identical to the one you designed in an earlier lab. By choosing a 10k resistor and a 150 pF capacitor, a clock frequency of 640 kHz is obtained.

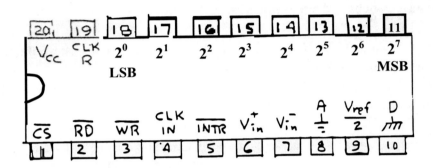

Figure 3. Pin Connections of the National Semiconductor ADC0804

The power supply for the 0804 may range from 2.5 to 6.5 volts. The chip has an analog ground pin (8) and a digital ground pin (10). Because digital signals tend to turn on and off very quickly at relatively high current levels compared to the analog signals, the digital ground tends to be quite noisy, so it is important to keep the digital connections connected to the digital ground and the analog connections connected to the analog ground.

Accurate A/D conversions depend on very accurate reference voltages. The 0804 is capable of producing its own reference voltage from the supply voltage or of operating with an externally applied reference voltage applied to pin 9. With no external connection to pin 9, the reference voltage is the power supply voltage. If exactly 5.12 volts is used as the power supply and if no external connections are made to pin 9, then V_{ref} = 5.12 volts and the LSB is 20 mV. That is, the measurement interval is broken down into 256 20 mV intervals, since 256 x 20 = 5120 mV = 5.12 volts. To produce a LSB smaller than 20 mV, it is necessary to use a different supply voltage or, preferably, to connect an external value of $V_{ref}/2$ = 128 x (desired LSB interval).

The range of the differential input voltage should be limited to $\mathbf{V_{ref}}$, since $\mathbf{V_{ref}}$ determines the LSB, the analog input voltage is thus represented at the output as a binary multiple of the LSB. For example, Suppose V_{ref} = 5.12 volts, and the analog input voltage is 1.20 volts. The LSB is 20 mV, so 1.2 volts represents 60 x the LSB voltage. Thus, the binary output of the ADC will be binary 60, which is 00111100.

What analog input voltage would a binary output of **01001101** represent if V_{ref} = 2.56 volts?

198

Pins 11 through 18 are the digital output, with pin 18 as the LSB and pin 11 as the MSB. These outputs are tri-state outputs, which means that the outputs appear as high impedances until they are enabled by an output enable signal. When an output enable signal is applied, the outputs appear as voltage sources which output either 0 volts or the power supply voltage. The power supply voltage at a digital output represents digital "1" and 0 volts at a digital output represents digital "0". Maximum current out of any digital output when it is at digital "1" is 6 mA (source current) and maximum current into any digital output when it is at digital "0" is 16 mA (sink current).

To enable the digital outputs, a digital "0" to pin 2 (RD) of the ADC must be applied. In this experiment, there is no reason to disable the digital outputs, so pin 2 can be connected to digital ground and left connected except when directed to do otherwise.

For a conversion to begin, it is necessary to simultaneously apply a digital "0" to pin 1 (CS) and 3 (WR). As soon as one of these pins is returned to digital "1", the conversion process will start. For testing purposes, it is satisfactory to leave pin 1 grounded to digital ground all the time. Pin 1 is needed when the ADC communicates directly with a microprocessor.

The conversion begins between 1 and 8 clock periods after pin 3 is returned to digital "1" assuming pin 1 remains at "0". First $V_{in}(+)$ is sampled, followed 4.5 clock periods later by $V_{in}(-)$. The MSB is tested first by a process of 8 comparisons. Lower order bits follow in order, each taking 8 comparisons to determine whether they should be "1" or "0". Thus, after 64 clock cycles, the conversion is complete. For a clock frequency of 640 kHz, this means that a conversion will take approximately 0.1 ms. Faster conversion times can be achieved with faster clock speeds, but accuracy is degraded at faster clock speeds.

At the completion of a conversion, pin 5 (INTR) goes from digital "1" to digital "0". This transition can be used as a control signal to let external circuitry know that the ADC is ready to provide new data. It can also be used to apply to pin 3 to start the conversion process over again so the ADC will operate in a "free running" mode or to enable the digital outputs of the ADC.

You are now ready to test an ADC. Try the following experiment, but first BE AWARE THAT THE 0804 IS A CMOS DEVICE. THIS MEANS THAT IT IS VERY SENSITIVE TO STATIC ELECTRICITY. IF IT COMES IN CONTACT WITH STATIC ELECTRIC CHARGE, SUCH AS ON CERTAIN CLOTHING, IT MAY BE PERMANENTLY DAMAGED. IT IS ADVISABLE TO BE HOLDING A LEAD CONNECTED TO GROUND WHEN HANDLING THE ADC. ONCE THE ADC IS PLUGGED INTO THE BREAD BOARD, IT IS RELATIVELY SAFE, BUT IS SHOULD NOT BE RUBBED WITH ANYTHING WHICH MIGHT CHARGE ITS CASE ELECTRICALLY.

THE EXPERIMENT PROCEDURE

1) Connect the 0804 as shown in Figure 4. Define one set of connections as analog ground and another set as digital ground. Run separate wires from the power supply ground connection to each of the ground connections on the breadboard. Note the circuit notations for analog and digital ground and be sure that components are connected to the appropriate ground point in the circuit. The 0.1 μF capacitors from pins 6 and 9 to analog ground are noise suppressors, as is the 10 μF electrolytic capacitor from pin 20 to digital ground. The 150 pF capacitor from pin 4 to digital ground establishes the internal clock frequency along with the 10 k resistor from pin 19 to pin 4. The 1.0 k resistors in series with the LEDs (Light Emitting Diodes) are current limiting resistors. The switch from pin 3 to ground can be a wire which is momentarily touched to the circuit digital ground connection. Be careful with the bare leads on the 8 sets of resistor/LEDs from the digital outputs to Vcc so they do not short circuit. It is best to run them perpendicular to the chip. Make sure the LED polarity is correct.

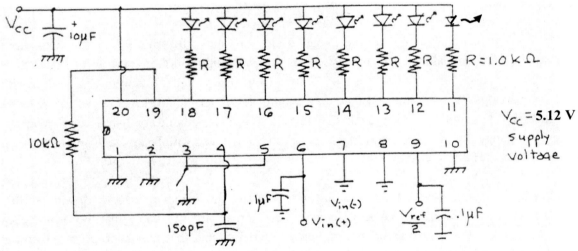

Figure 4.   0804 ADC Test Circuit

2) Adjust the voltage from pin 20 to ground to as close to 5.120 volts as you can get. Measure this voltage with the Keithley meter. Record the value you end up with. Note that because of the type of control in the power supply, you may need to settle for a slightly different voltage.

3) Measure the voltage from pin 9 to analog ground. It should be exactly half the voltage on pin 20, and hopefully, 2.560 volts. Record your exact value.

4) Now measure the internal clock frequency and waveform to be sure the clock is working. Use your CRO and your frequency counter. The frequency should be 640 kHz, but it usually ends up closer to 300 kHz. The exact frequency should be $f = 1 / (1.1RC)$.